POLARIZATION SPECTROSCOPY OF IONIZED GASES

ASTROPHYSICS AND SPACE SCIENCE LIBRARY

VOLUME 200

POLARIZATION SPECTROSCOPY OF IONIZED GASES

by

S. A. KAZANTSEV

Institute of Physics,
St. Petersburg State University, Russia

and

J.-C. HÉNOUX

Observatoire de Paris,
Meudon, France

SPRINGER-SCIENCE+BUSINESS MEDIA, B.V.

A C.I.P. Catalogue record for this book is available from the Library of Congress

ISBN 978-90-481-4550-8 ISBN 978-94-017-2708-2 (eBook)
DOI 10.1007/978-94-017-2708-2

Cover: Electrons and protons accelerated above the surface of the Sun in the corona are precipitating on the solar atmosphere. As a result the atmosphere is heated and a hot ionized plasma rises in the corona. Particle acceleration takes place in regions where magnetic fields are present and the ionized hydrogen plasma is then trapped in gigantic magnetic loops. When it cools, the ionized hydrogen recombines with local electrons, becomes neutral and starts emitting in the Red Balmer α line. At the base of the loop, the accelerated protons and electrons are bombarding the atmospheric neutral hydrogen leading to linearly polarized Balmer α line emission. The measurement of this polarization provides information on the nature and velocity distribution of the accelerated particles. Courtesy of Z. Mouradian, DASOP, URA326, Observatoire de Paris.

Printed on acid-free paper

CONTENTS

PREFACE

This book describes the physical principles of polarization spectroscopy and its applications to the remote sensing of ionized gases. Recent evolution of this technique allows for quantitative studies of energy transport and dissipation in various types of ionized gases states.

In the theoretical part, the basic phenomena of the ordering of the velocities of fast exciting charged particles, together with the polarization of the outer electron shells of the ensemble of excited atoms or molecules are described. A general approach based on the irreducible tensorial set representation of the rotation symmetry group is used. The effects of the polarization of the excited atoms or molecules are examined in more detail. Then the integral equation giving the intensity and polarization of the emitted lines is derived and methods to solve this equation are analysed.

Experimental applications of remote sensing are reviewed. Universal spectropolarimetric remote sensing has been applied to laboratory low pressure gas discharge plasmas and to non-thermal processes taking place in the solar atmosphere, illustrating the possibilities of this new method.

This book may be useful for researchers, Ph D students and graduate students utilizing optical methods for the remote sensing of various ionized gases: low temperature plasmas generated in different discharges and beam-gas systems, high temperature plasmas, solar plasmas, eruptive processes, ionized gases in the upper atmosphere of the earth where precipitating particles are present and various other cases.

INTRODUCTION

Elaborate remote sensing of ionized gases is based on the study of their optical line spectrum. Whereas the conventional intensity spectroscopic methods already attained the limits of their possibilities, polarization spectroscopy represents by itself a new optical diagnostic technique which expands considerably the frontiers of the studies of ionized gases by optical methods.

Since the discovery of splitting of atomic lines in polarized components by Zeeman /1/, the main application of polarization spectroscopy, as a remote sensing technique, has been the measurement of the magnetic field strength and orientation in astrophysical objects. These measurements make use of the direct connection between the local vector magnetic field and the observed spectroscopic polarization pecularities. In astronomy, the application of the spectropolarimetry to magnetic field measurements goes back to 1908 when Hale /2/ discovered the magnetic field of sunspots. In locations were the magnetic field is low, like in solar prominences, it can be measured using Hanle effect /3 , 4/. The achievements of the very active field of Zeeman polarimetry in solar physics are presented in detail in J.O. Stenflo's /5/.

Polarized radiation can also be generated even in the absence of a magnetic field, and polarization measurements can bring information on plasma anisotropies providing new applications of the polarimetric remote sensing. These possibilities in the optical spectral range and in X-rays have been proposed and used /6, 7, 8/ as a method of sensing non-thermal processes in the solar atmosphere. The implementation of this spectropolarimetric sensing requires the utilization of reliable atomic data and of advanced methods of theoritical atomic physics /9/.

Restrospectively, up until the last decade, there was a lack of general theoretical relationships between the polarization spectroscopic parameters and the kinetic characteristics of the exciting particles. Therefore spectropolarimetry of ionized gases was not a quantitative tool. Due to the complexity of the spectropolarimetric effects, the absence of specific atomic constants and cross-sections, and the difficulties encountered in measuring the Stokes parameters in the line spectrum of ionized gases, impact spectropolarimetric effects were not used for the remote sensing of plasma parameters. The history of impact polarization, however, began in 1926 with the detection of the linear polarization of some spectral lines of mercury in a low pressure arc by Skinner /10/. Although the polarization of spectral lines has since been observed in different discharges (positive column of D.C. discharges, hollow cathode discharges, high frequency discharges, etc), only qualitative analysis of the discharge objects were made during those rare spectropolarimetric studies.

Polarization of emission and absorption lines is related to the unequal distribution of populations of the magnetic substates of excited atoms, in other words to the ordering of momenta of the outer electron shells of atomic particles (atoms, molecules, ions). From a number of detailed experimental studies it was concluded that for a majority of cases the polarization of the optical line emission of ionized gases is determined by the self-alignment of atomic particles or the quadrupole orientation of momenta of electron shells induced in the excitation process. Different fundamental physical mechanisms, such as reabsorption of radiation, anisotropic electron or proton impact excitation and anisotropy of the collisional relaxation are responsible for the formation of the self-alignment. In this respect the self-alignment of atomic ensembles may be regarded as a general feature reflecting the limited spatial dimensions and the energy exchange processes in any ionized gas entity /11/.

The most promising diagnostic application is the use of the self-alignment induced by the anisotropy of the motion of fast exciting electrons. The energy delivered into an ionized gas and the internal electric field distribution lead to a localised anisotropy of the velocity distribution function of fast electrons. Then the resulting anisotropy in the electron velocity space gives rise to the collisional self-alignment of the excited atoms and to the polarization of the optical emission. Later on, the transformation by collisons of the energy of fast electrons into a thermal form diminishes the degree of anisotropy in velocity space and consequently the degree of linear polarization. Qualitative studies of the polarization degree in the spectra of gas discharges that results from electron impact excitation of atoms have shown the extensive informative possibilities of spectropolarimetric techniques. New theoretical and experimental studies of the polarization effects in ionized gases under electron impact excitation and of the derivation of plasma parameters enable one to regard the polarization spectroscopy as a new and quantitative remote spectral sensing technique.

The principally new potential of polarization spectroscopy results from the general relationship between the alignment tensor of the excited atomic ensemble at the collisional excitation and the quadrupole moment of the velocity distribution function of fast electrons /12/. By nature the quadrupole moment is the tensor of anisotropic pressure of the free electron gas and is proportional to the tensor of the momentum flow of electrons and thereby takes into account the spatial variation of the energy exchange. By analysing the spatial distribution of the spectropolarimetric parameters in different spectral lines it is possible to determine the components of the anisotropic pressure tensor of electrons in different parts of the ionized gas. It is thus possible to estimate the relative contribution of these regions to the energy balance into the ionized gas entity. By these means spectropolarimetric techniques offer a unique method for remote sensing of the structural characteristics of an ionized gas entity that is beyond the limits of ordinary spectroscopic sensing.

The needs of diagnostics are present in many fields of physics, astrophysics and geophysics. Charged and neutral particles with anisotropic velocity distribution functions are present in plasma-beam systems, pulsed discharges and tokomaks. In the Sun and in active stars, particles are accelerated during solar and stellar flares. By the use of spectropolarimetric techniques, low energy proton beams, practically undetected until now, have been shown to be present in solar flares, and to contribute significantly to energy transport and dissipation. The interaction of solar wind with planetary magnetospheres and the precipitation of charged particles in polar aurorae could also generate polarized line emission. In all cases the measurement of the degree of polarization and of its orientation will give a clue to the understanding of the energy transport mechanism acting in the laboratory, geophysical or astrophysical plasma under study.

This book represents the first detailed description of the principles of the polarization spectroscopy and of the remote sensing of ionized gases by spectropolarimetry. The application of this technique in X-rays was shown to be very efficient for the study of the anisotropies in hot laboratory plasmas. In astronomy, its use has been restrained only by the difficulty to achieve an appropriate space resolution at these short wavelengths. The applications presented in this book are restrained to the visible and UV part of the optical spectrum. Much attention is being given to the phenomenon of self-alignment of excited atoms induced by collisions and to the new possibilities it offers in the determination of atomic constants. The power of the spectropolarization diagnostic technique is illustrated by the determination of some structural characteristics of laboratory gas discharges and of solar flare plasmas.

The first chapter represents the concise quantum mechanical description of an atomic ensemble polarization and the polarization effects on the line emission spectrum of an ionized gas. The second is devoted to the principles of the spectropolarimetric method. The contents of these chapters requires the reader to be familiar enough with the corresponding parts of the quantum mechanics of atomic systems. The desire to maintain the integrity of the book led to the high degree of brevity in the description of the quantum mechanical apparatus. The authors considered such a style of description acceptable, keeping in mind that the general theory of polarization effects in atomic spectroscopy is substantionally given in the well-known book "Density matrix theory and its applications" by K.Blum /13/. The remaining chapters are dedicated to applications of the polarization spectroscopy in ionized gases of different physical nature and atomic spectroscopy. Much attention is paid to laboratory gas discharges which on the one hand perfectly illustrate the peculiarities of polarization effects and the general potential of the technique to the diagnostic of the physical conditions in low temperature plasma and on the other hand could simulate, to a certain degree, polarization spectroscopic effects in astrophysics and geophysics. Chapter 7 is devoted completely to these effects in astrophysical objects, focusing mainly on the processes in the solar atmosphere, where the first quantitative results have been recently obtained. The main contribution to chapters 1-6 have been made by S.A.Kazantsev and to chapter 7 by J.-C.Hénoux.

CHAPTER 1

SPECTROPOLARIMETRIC MANIFESTATION OF SELF-ALIGNMENT

The formation of the quadrupole ordering of angular momenta of electron shells of excited atomic particles (self-alignment of atoms or ions) in ionized gas and its spectroscopic manifestations are discussed below. This effect is analysed on the basis of the symmetry of the elementary excitation processes. For the most simple case of a low temperature plasma typical for laboratory gas discharges, these are self photo-excitation and inelastic collisions with electrons. The spectropolarimetric plasma diagnostic is based on the collisional self-alignment of atomic particles and on the property of the polarization of excited states induced by electron-atom collisions, to be connected to the quadrupole moment of the velocity distribution function of fast electrons.

Spectroscopically, the self-alignment manifests itself in the polarization of the collisionally excited spectral lines. This phenomenum is described theoretically and expressions for the Stokes parameters of the optical emission in terms of the kinetic characteristics of exciting electrons are given in this chapter. These expressions are used to examine the diagnostic possibilities of this method. Since a fundamental question for the implementation of the spectropolarimetric diagnostic is the availability of reliable data on the electron impact alignment cross-sections, which are the kernels of the basic integral equations of the polarization spectroscopy, the determination of the cross-sections is discussed in this chapter as well.

1.1. Representation of the Stokes parameters by polarization moments

In order to examine the spectroscopic manifestation of self-alignment of an ensemble of excited atomic particles in an ionized gas, let us regard the classical concept and the theoretical quantum mechanic means for the description of the polarization of this ensemble. First, limiting ourselves to a qualitative description, we represent the polarization of an ensemble of atomic particles in terms of the spatial anisotropy of angular momenta distribution. In the first stage we shall discuss the model of the linear oscillator in order to describe the polarization phenomena.

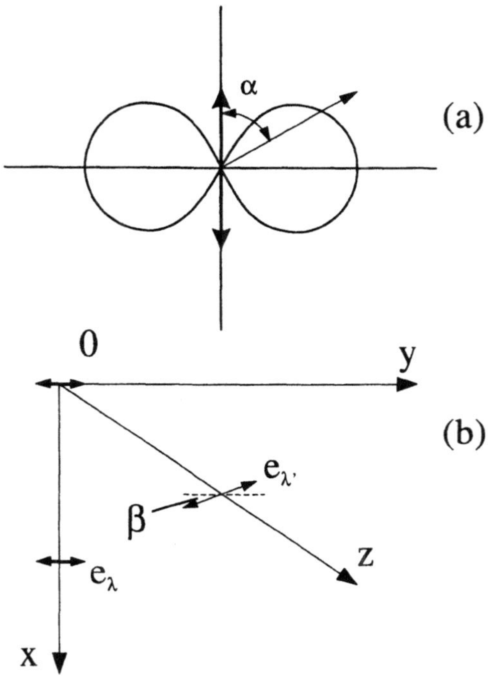

Figure 1. **Spatial diagram of the radiation from an electric dipole (a) and representation of the coordinate system used for the classical description of self-alignment and Hanle effect (b)**

1.1.1 CLASSICAL AND POLARIZATION MOMENT FORMALISM

In order to describe the polarization of an excited atomic ensemble by means of the quantum mechanical formalism of polarization moments, it is convenient to begin with the classical model of an atom as a linear oscillator formed by a heavy positive core and an outer electron, which is undergoing a forced oscillation under the influence of an incident electromagnetic wave. This model has been used widely since the days of Hanle's work for a graphic interpretation of several optical phenomena associated with the interaction of resonant radiation with an ensemble of particles in an external magnetic field /3/. Hanle studied the depolarization of the emission from mercury vapor excited by polarized resonant light in an external magnetic field, and he suggested a simple interpretation for his observations based on the model of a damped electric dipole which is precessing in an external magnetic field. The radiation from a linear harmonic oscillator in the wave zone is represented by spherical electromagnetic waves, and the angular dependence of the radiation intensity is described by

$$J = J_0 \sin^2 \alpha$$

where α is the angle between the observation direction and the oscillation axis of the oscillator (Fig. 1).

Let us consider an ensemble of free atoms at the origin of coordinates, being excited by a beam of linearly polarized light. The light beam is directed parallel to the OX axis, and the polarization vector of the light, $\vec{e}_\lambda$, is parallel to the OY axis (Fig. 1). The linear oscillators with axes parallel to $\vec{e}_y$ are excited, and their radiation in the OZ direction is linearly polarized with a polarization vector parallel to OY. We now impose on this ensemble of atoms an external magnetic field directed along the OZ observation axis, and we orient the axis of the analyzer in the observation channel at an angle β from the OY axis. If the linear oscillator is excited at the time $t = t_0$, then the time evolution of its radiation intensity is described by

$$dJ = J_0 \exp\{-\Gamma(t - t_0)\} \cos^2\{\omega(t - t_0) - \beta\};$$

where $\Gamma = 1/\tau$ is the decay constant of the oscillator, τ is the lifetime of the excited atom, $\omega = \gamma \cdot H$ is the Larmor frequency of the precession, γ is the gyromagnetic ratio, and H is the external magnetic field.

The intensity of the measured polarized radiation from the entire ensemble of atoms is given by the integral

$$J = J_0 \int_{-\infty}^{1} dt_0 \exp\{-\Gamma(t - t_0)\} \cos^2\{\omega(t - t_0) - \beta\}$$

After integrating we find

$$J = \frac{J_0}{2\Gamma}\left[1 + \frac{\cos 2\beta - x \sin 2\beta}{1 + x^2}\right] = \frac{J_0}{2\Gamma}\left[1 + \frac{\cos(2\beta + \arctan x)}{\sqrt{1 + x^2}}\right] \qquad (1.1)$$

where $x = 2\omega\tau = 2\gamma H/\Gamma$

Expression (1.1) determines the intensity J of the polarized resonant radiation from the ensemble of atoms as a function of the external magnetic field H and of the orientation of the linear analyser; i.e., it gives the shape of the Hanle-effect signal. It follows from (1.1) that the measured intensity reaches a maximum when the angle β is

$$\beta = -(1/2)\arctan(2\omega/\Gamma) = -(1/2)\arctan(2\omega\tau).$$

This means that an external magnetic field rotates the direction of the predominant oscillations of the electric vector of the scattered resonant radiation in the coordinate system selected in the XOY plane, and the rotation angle can be determined by varying the orientation of the axis of the analyzer in the observation channel until the detected intensity is maximized. It can be seen from (1.1) that the maximum intensity falls off with increasing magnetic field. In a strong field, $\omega\tau \gg 1$, the oscillator will precess at a frequency so high that it performs several rotations in the XOY plane, J becomes independent of β and the radiation detected under these circumstances is depolarized.

We now excite these atoms with a beam of resonant unpolarized light, again following the OX direction. The oscillation of the vector E may be thought of as two mutually orthogonal, uncorrelated oscillations along the OY and OZ axes. These orthogonally polarized waves will excite two incoherent oscillators with oscillation axes parallel to OY and OZ, respectively. Along the OZ observation direction we will again detect linearly polarized emission with a polarization vector parallel to the OY axis, since in this direction the oscillator with oscillation axis parallel to OZ does not radiate (Fig. 1). An external magnetic field applied along the OZ observation axis does not perturb the oscillations of an oscillator with axis parallel to OZ. The change in the polarization characteristics of the radiation detected is due to the precession of an oscillator with an original oscillation axis along OY in the XOY plane, as in the case of the linearly polarized beam of incident light. In the case of excitation by unpolarized light, the qualitative behavior of the polarization of the resonant radiation is thus described by expression (1.1).

Using a purely quantum mechanical description, the state of an excited atomic ensemble may be theoretically described in terms of the statistical operator $\hat{\rho}$ with the kernel given by /14/

$$\rho(x, x') = \overline{\Psi(x)\Psi^*(x')}$$

where $\Psi(x)$ is the wave function of the electron shell of a single atom or ion, x is the spatial and spin coordinates, and dash stands for the averaging over the ensemble. The following integral transformation represents the operation of the statistical operator $\hat{\rho}$ on an atomic function $\varphi(x)$:

$$\hat{\rho}\varphi = \int dx' \rho(xx')\varphi(x')$$

Here the integration over the spatial variables and summation over the spin coordinates are assumed.

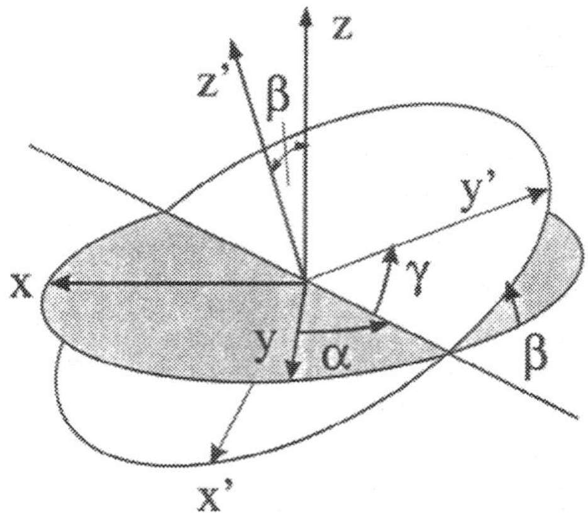

Figure 2. Euler angles used in the transformation from frame of reference (XYZ) to frame of reference (X'Y'Z')

In the first stage we restrict ourselves to the consideration of the ionized gas of even isotopes of atoms with well-resolved fine structure and no hyperfine structure (for example, noble gases). With the help of a spectroscopic instrument with medium resolution it is possible to extract spectral lines corresponding to optically allowed transitions between atomic states. These states without external perturbation are degenerated for the magnetic quantum numbers m. In the space of atomic states it is natural to use eigenfunctions of the momentum operator of electronic shell $\mid \delta, J, m\rangle$ as the basic set (J, m represent the angular momentum and its projection on the quantization axis for the quantum state defined by the quantum number δ).

Functions $\mid \delta, J, m\rangle$ form the basis of the irreducible representation of the group of rotation in the space of atomic functions /15/. These representations are implemented by the Wigner D-matrix $D^{(\kappa)}_{qq'}$ which perform the rotational transformations of $\mid \delta, J, m\rangle$ functions under the rotation of the frame of reference /16/:

$$\mid \delta, J, m'\rangle = \sum_{m} D^{(J)}_{mm'}(\alpha, \beta, \gamma) \mid \delta, J, m\rangle$$

here (α, β, γ) are Euler angles (Fig. 2) defining the rotation $(x, y, z) \rightarrow (x', v', z')$*and* $\mid J, m'\rangle$ is the wave function in the final reference frame. Using the expression of a wave function as the superposition of eigenstates $\mid \delta, J, m\rangle$:

$$\mid \Psi >= \sum_{\delta, J, m} C_{\delta, J, m} \mid \delta, J, m\rangle,$$

($C_{\delta,J,m}$ are complex coefficients) one may write for the statistical operator:

$$\hat{\rho} = \overline{\mid \Psi\rangle\langle\Psi \mid} = \sum_{\delta,J,m,\delta',J',m'} \mid B\delta', J', m'\rangle \rho_{\delta',J',m',\delta,J,m} \langle \delta, J, m \mid \qquad (1.2)$$

where

$$\rho_{\delta' J' m',\delta,J,m} = \overline{C_{\delta',J',m'} C^*_{\delta,J,m}}$$

is known as the density matrix in $J - m$ representation /16, 17/.

For the purpose of physical clearness of the spectropolarimetric phenomena we shall use the basis of the orthonormal irreducible tensorial operators /18, 17, 19/ made from the eigen functions of the angular momentum $\mid J, m\rangle$ /13/:

$$\hat{T}_q^{(\kappa)} = \sum_{m,m'} (-1)^{J-m} \begin{bmatrix} J' & J & \kappa \\ m' & -m & q \end{bmatrix} \mid J'm'\rangle\langle Jm \mid \qquad (1.3)$$

where $\begin{bmatrix} J' & J & \kappa \\ m' & -m & q \end{bmatrix}$ is the Clebsh-Gordan coefficient.

The statistical operator $\hat{\rho}$ then may be expanded in terms of the irreducible tensorial operators $\hat{T}_q^{\kappa}$:

$$\hat{\rho} = \sum_{\kappa,q,J,J'} Sp\left[\hat{\rho} T_q^{(\kappa)+}(J', J)\right] T_q^{(\kappa)+}(J, J') = \sum_{\kappa,q,J,J'} \rho_q^{(\kappa)} T_q^{(\kappa)+}(J', J) \quad (1.4)$$

Here

$$\rho_q^{(\kappa)}(J', J) = \sum_{m,m'} (-1)^{J'-m'} (2\kappa + 1)^{1/2} \begin{pmatrix} J' & J & \kappa \\ m' & -m & q \end{pmatrix} \rho_{J',m',J,m}$$

are the multipoles of the state /13/, Sp is the sum of diagonal elements.

For the single state under study ($J' = J$), $\rho_q^{(\kappa)}(J, J) = \rho_q^{(\kappa)}$ is known as polarization moments of atomic density matrix /17, 18/. Rotational transformation of the polarization moments is given by Wigner D-matrix /16/:

$$^{z'}\rho_{q'}^{(\kappa)} = \sum_{q} D_{qq'}^{(\kappa)^*}(zz')\, {}^{z}\rho_q^{(\kappa)}. \qquad (1.5)$$

Description of the atomic ensemble by means of the polarization moment formalism takes into consideration the angular symmetry of a system. The macroscopic ordering of the spatial distribution of angular momenta results in polarization features of the optical emission /17, 18/. Zero order moment $\rho_0^{(0)}$ is scalar, proportional to the population of the state. The three elements $\rho^{(1)}_{0,\pm1}$ may be regarded as the cyclic components of a certain vector known as the orientation vector. This vector defines the dipole ordering of the excited state angular momenta. Five components, $\rho^{(2)}_{0,\pm1,\pm2}$, form the alignment tensor, determining the quadrupole ordering of angular momenta of atomic particles. Component $\rho^{(2)}_0$ is called the longitudial alignment. It is expressed as:

$$\rho^{(2)}_0 = \sqrt{5}\,[(2J+3)(2J+1)(J+1)J(2J-1)]^{-1/2}$$

$$\sum_m \left[3m^2 - J(J+1)\right] \rho_{JmJm}. \qquad (1.6)$$

1.1.2 RELATION TO THE ORDERING OF ANGULAR MOMENTA

Description of the polarization characteristics of the optical field may be realized with the help of the expansion of the photon states using the basis of the irreducible representation of the rotation group which is formed by the states with the right and left circular polarizations $\mid \pm1\rangle$ and longitudinal polarization $\mid 0\rangle$ /20,13, 17/. The state of the ensemble of photons which are emitted by the macroscopic small atomic ensemble in the ionized gas on the atomic transition $J_1 \rightarrow J$ is represented as

$$\mid \varepsilon\rangle = a_0 \mid 0\rangle + a_1 \mid 1\rangle + a_{-1} \mid -1\rangle,$$

where $a_{0,\pm1}$ are the probability amplitudes. The longitudinal component a_0 appears only if the quantization axis of the reference frame is not parallel to the direction of the light propagation.

The eigen-vectors of the photon states are transformed the same way as the eigen-states of the angular momentum with $J = 1$:

$$\mid \mu'\rangle = \sum_\mu D^{(1)}_{\mu m u'}(\alpha,\beta,\gamma) \mid \mu\rangle$$

where $\mid \mu\rangle$ and $\mid \mu'\rangle$ are the basic photon states in the initial and rotated coordinate frame.

The ensemble of photons emitted by the ensemble of atomic particles on the $J_1 \rightarrow J$ transition for a certain time interval (Fig. 3) may be described with the help of the statistical operator $\hat{V} = \overline{\mid \varepsilon\rangle\langle\varepsilon \mid}$ (dash means the ensemble averaging):

$$\hat{V} = \sum_{\mu\mu'} \mid \mu'\rangle \overline{a_{\mu'}a^*_\mu} \langle\mu \mid = \sum_{\mu\mu'} \mid \mu'\rangle V_{\mu\mu'} \langle\mu \mid \qquad (1.7)$$

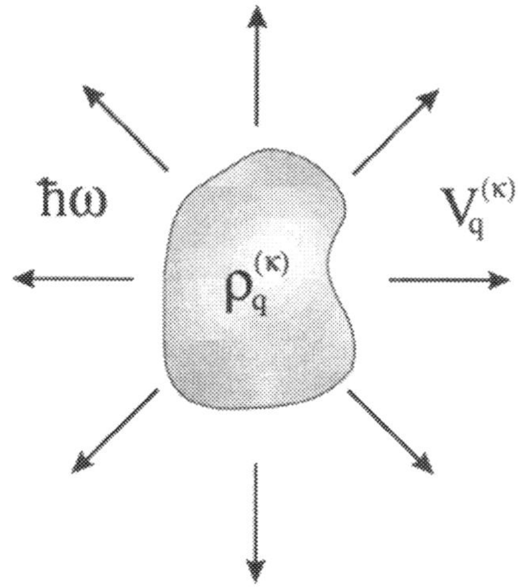

Figure 3. Connection between the polarization moments of optical radiation and the polarization moments of an atomic ensemble

The set $V_{\mu\mu'}$ plays the role of the polarization density matrix of photons. Similar to the atomic statistical operator, the statistic operator of photons may be expanded in terms of the irreducible tensor operators $\hat{W}_q^{(\kappa)}$ which are built up from the basic photon states as follows /21, 19/:

$$\hat{W}_q^{(\kappa)} = \sum (-1)^{\mu'+1}\sqrt{2\kappa+1}\begin{pmatrix} 1 & 1 & \kappa \\ \mu' & -\mu & -q \end{pmatrix} \mid \mu'\rangle\langle\mu \mid \qquad (1.8)$$

The expansion may then be written as:

$$\hat{V} = \sum_{\kappa,q} V_q^{(\kappa)} \hat{W}_q^{(\kappa)}$$

where

$$V_q^{(\kappa)} = \sum_{\mu\mu'} (-1)^{1-\mu'}\sqrt{2\kappa+1}\begin{pmatrix} 1 & 1 & \kappa \\ \mu' & -\mu & -q \end{pmatrix} V_{\mu'\mu} \qquad (1.9)$$

are the polarization moments of the photon operator. Operators $\hat{W}_q^{(\kappa)}$ and the polarization moments $V_q^{(\kappa)}$ are transformed under the coordinate system rotation by the same formulae as (1.5).

In order to construct the density matrix of photons in a nonrelativistic approach, let us use the connection between the probability amplitude of the photon emerging with μ - polarization resulting from the atomic transition $\mid J_1, m_1\rangle \rightarrow \mid J, m\rangle$ and the dipole moment $\langle Jm \mid \hat{d}_\mu \mid J_1, m_1\rangle$. Probability amplitude for the photon emission in the direction $\vec{n}$ due to the

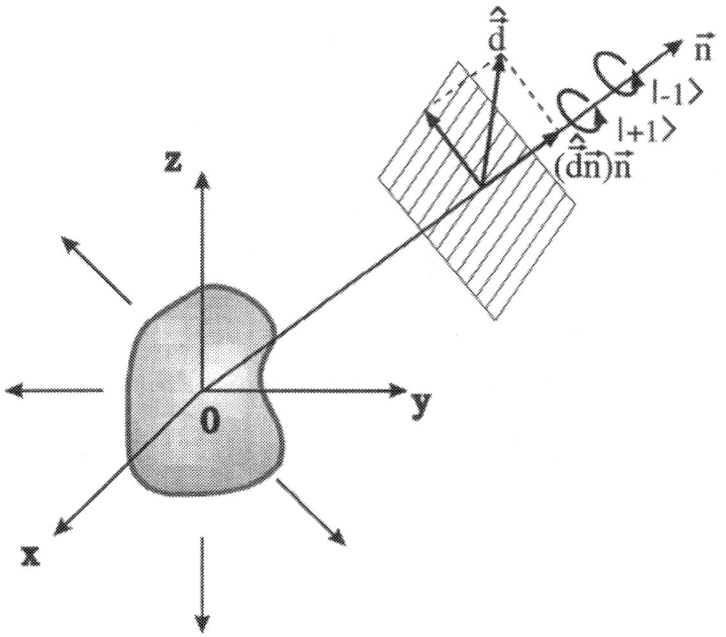

Figure 4. Projection of the vector-operator of the dipole moment on the plane orthogonal to the direction of the light beam

transversality of the light wave is proportional to the cyclic components of the operator $\left(\hat{\vec{d}} - (\hat{\vec{d}}\vec{n})\vec{n}\right)$ (Fig. 4):

$$\left(\hat{\vec{d}} - (\hat{\vec{d}}\vec{n})\vec{n}\right) = \sum_{\nu}(\delta_{\nu\mu} - n^*_\nu n_\mu)\hat{d}_\nu \tag{1.10}$$

where $\hat{d}_\nu$ and n_ν are cyclic components of the vector operators $\vec{d}$ and $\vec{n}$. For the usual system of coordinates with the quantization axis parallel to $\vec{n}$ the transversality condition is fulfilled, so

$$\left(\hat{\vec{d}} - (\hat{\vec{d}}\vec{n})\vec{n}\right)_{\pm 1} = \hat{d}_{\pm 1}$$

and

$$\left(\hat{\vec{d}} - (\hat{\vec{d}}\vec{n})\vec{n}\right)_0 = 0.$$

The density matrix of photons emitted by an atomic ensemble per second as a result of the quantum transitions $J_1 - J$, averaged over the subsystem of the lower level J_1, /13/ will take the form:

$$V_{\mu'\mu} = B(\omega)N_a \int_{4\pi} d\vec{n} \sum_{m_1,m,m'} \langle J_1 m_1 \mid \left(\hat{\vec{d}} - (\hat{\vec{d}}\vec{n})\vec{n}\right)_{\mu'} \mid Jm\rangle$$

$$\langle Jm \mid \hat{\rho}_J \mid Jm'\rangle\langle Jm' \mid \left(\hat{\vec{d}} - (\hat{\vec{d}}\vec{n})\vec{n}\right)^{+}_{\mu} \mid J_1 m_1\rangle \qquad (1.11)$$

where N_a is the total number density of atomic particles, ω the frequency of the optical transition, $B(\omega) = \omega^4/2\pi c^3$, and c the velocity of light.

The expression (1.11) may be transformed as:

$$V_{\mu'\mu} = \sum_{\nu\nu'} P_{\mu'\nu'\mu\nu}(V_0)_{\nu'\nu} \qquad (1.12)$$

where

$$(V_0)_{\nu'\nu} = 4\pi B(\omega)N_a$$

$$\sum_{m_1 m m'} \langle J_1 m_1 \mid d_{\nu'} \mid Jm\rangle\langle Jm \mid \hat{\rho}_J \mid Jm'\rangle\langle Jm' \mid d^{+}_{\nu} \mid J_1 m_1\rangle \qquad (1.13)$$

and the tensor

$$P_{\mu'\nu'\mu\nu} = \frac{1}{4\pi}\int_{4\pi} d\vec{n}\left(\delta_{\nu'\mu'} - n^*_{\nu'}n_{\mu'}\right)\left(\delta_{\nu\mu} - n_\nu n^*_\mu\right) \qquad (1.14)$$

accounts for the transversality of the electromagnetic wave.

Making use of the fact that the cyclic components of the unit vector are the spherical harmonics of the first rank : $n_\nu = C^{(1)}_\nu(\vec{n})$ and of the expansion of the products of the spherical functions /16/ :

$$n_\nu n^*_{\nu'} = (-1)^{\nu'} C^{(1)}_{\nu} C^{(1)}_{-\nu} = \sum_{Lm} (-1)^{\nu'} \begin{bmatrix} 1 & 1 & L \\ 0 & 0 & 0 \end{bmatrix} \begin{bmatrix} 1 & 1 & L \\ \nu & -\nu' & m \end{bmatrix} C^L_m(\vec{n})$$

we have:

$$P_{\mu'\nu'\mu\nu} = \frac{1}{3}\delta_{\nu'\mu'}\delta_{\nu\mu} + \sum_{Lm} \frac{(-1)^{\nu'+\mu'}}{2L+1}$$

$$\begin{bmatrix} 1 & 1 & L \\ 0 & 0 & 0 \end{bmatrix}^2 \begin{bmatrix} 1 & 1 & L \\ \nu' & -\nu & m \end{bmatrix} \begin{bmatrix} 1 & 1 & L \\ \mu' & -\mu & m \end{bmatrix} \qquad (1.15)$$

Polarization moments of the density matrix of photon ensemble will take the form:

$$V^{(\kappa)}_q = \left\{\frac{1}{3} + (2\kappa+1)^{-1} \begin{bmatrix} 1 & 1 & \kappa \\ 0 & 0 & 0 \end{bmatrix}\right\} (V_0)^{(\kappa)}_q \qquad (1.16)$$

where:

$$(V_0)_q^{(\kappa)} = \sum_{\nu\nu'}(-1)^{1-\nu'}\sqrt{2\kappa+1}\begin{pmatrix} 1 & 1 & \kappa \\ \nu' & -\nu & -q \end{pmatrix}(V_0)_{\nu'\nu} \tag{1.17}$$

The components of this tensor are the same as in /22/:

$$V_0^{(0)} = 2\,(V_0)_0^{(0)}/3, V_q^{(1)} = (V_0)_q^{(1)}/3, V_q^{(2)} = 7\,(V_0)_q^{(2)}/15$$

Let us now determine the components of the photon operator $(V_0)_q^{(\kappa)}$. Representing the statistical operator as the product of the irreducible tensor operators (1.4), (1.3) and making use of the Wigner-Eckart theorem, we have:

$$(V_0)_{\nu'\nu} = 4\pi B(\omega)N_a \sum_{\kappa,q}\sqrt{2\kappa+1}\rho_q^{(\kappa)} \mid \langle J_1 \parallel \hat{d} \parallel J\rangle \mid^2 \sum_{mm_1m'}(-1)^{2J_1+J-2m_1-m}$$

$$\begin{pmatrix} J_1 & 1 & J \\ -m_1 & \nu' & m \end{pmatrix}\begin{pmatrix} J & \kappa & J \\ -m & q & m' \end{pmatrix}\begin{pmatrix} J_1 & 1 & J \\ -m_1 & \nu & m' \end{pmatrix}$$

where $\langle J_1 \parallel \hat{d} \parallel J\rangle$ is the reduced matrix element of the dipole operator. After straightforward transformations for the density matrix of photon ensemble, we have:

$$(V_0)_{\nu'\nu} = 4\pi B(\omega)N_a \sum_{\kappa,q}\rho_q^{\kappa} \mid \langle J \parallel \hat{d} \parallel J_1\rangle \mid^2 (-1)^{J+J_1+\nu}\sqrt{2\kappa+1}$$

$$\begin{pmatrix} 1 & 1 & \kappa \\ \nu' & -\nu & q \end{pmatrix}\begin{Bmatrix} 1 & 1 & \kappa \\ J & J & J_1 \end{Bmatrix}$$

Using the definition of the irreducible tensor operators of the photon field, the last expression may be written in the following form:

$$\hat{V}_0 = \sum_{\nu'\nu} \mid \nu'\rangle(V_0)_{\nu'\nu}\langle\nu \mid = 4\pi B(\omega)N_a\sum_{\kappa q}\rho_q^{(\kappa)}(-1)^{J+J_1+q-1}$$

$$\mid \langle J \parallel \hat{d} \parallel J_1\rangle \mid^2 \begin{Bmatrix} 1 & 1 & \kappa \\ J & J & J_1 \end{Bmatrix}\hat{W}_{-q}^{(\kappa)} = \sum_{\kappa q}(V_0)_q^{(\kappa)}\,W_q^{(\kappa)} \tag{1.18}$$

This expression is the expansion of the statistical operator $\hat{V}_0$ in terms of the basis operators $W_q^{(\kappa)}$. Coefficients of this expansion are the polarization moments of the photon operator:

$$(V_0)_q^{(\kappa)} = 4\pi B(\omega)N_a(-1)^{J+J_1+q-1}\rho_{-q}^{(\kappa)} \mid \langle J \parallel \hat{d} \parallel J_1\rangle \mid \begin{Bmatrix} 1 & 1 & \kappa \\ J & J & J_1 \end{Bmatrix} \tag{1.19}$$

By these means, taking into consideration relationship (1.16), we see that the polarization moments of the statistical photon operator are proportional to the polarization moments of the density matrix of the excited state of the same rank κ . This important result may be considered the conservation of the type of symmetry or the transfer of the type of ordering of the angular momenta of the excited atomic ensemble into the polarization of the electromagnetic field emerging under spontaneous decay.

1.1.3 STOKES PARAMETERS WITHIN THE DETECTOR FRAME OF REFERENCE

The spectroscopic diagnostic of an ionized gase is based on the study of its own emission, that propagates within a small solid angle $\tilde{\tau}$ (Fig. 5). Examination of the polarization features of such a beam is more easy to carry on within a frame of reference with quantization axis oriented along the direction of the beam. This reference frame is known as the detector frame of reference. In this coordinate system, due to the transversality of the electromagnetic radiation, only the photon states $| \pm 1\rangle$ will be excited, and hence the density matrix will comprise only four non-zero components V_{11}, V_{-11}, V_{1-1}, V_{-1-1}. The set of these components known as the polarization density matrix of the light beam in the spiral representation $U_{\mu'\mu}$ /13, 20/ is expressed using (1.10), (1.11) as:

$$U_{\mu'\mu} = B(\omega)N_a$$

$$\tilde{\tau} \sum_{m_1 m m'} \langle J_1 m_1 \mid d_{\mu'} \mid Jm\rangle\langle Jm \mid \hat{\rho}_J \mid Jm'\rangle\langle Jm' \mid d_\mu^+ \mid J_1 m_1\rangle \tag{1.20}$$

Let the registration system detects the photons with the polarization defined by the vector $| \vec{e}\rangle$. In this case the probability or the photon detection in the state $| \varepsilon\rangle$ is given by $| \langle \vec{e} \mid \varepsilon\rangle |^2$ and the detected intensity, using (1.20) and /13/ will be transformed to:

$$I(\vec{e}) = \bar{|}\langle\vec{e} \mid \varepsilon\rangle |^2 = \sum_{\mu'\mu}\langle\vec{e} \mid \mu'\rangle U_{\mu'\mu}\langle\mu \mid \vec{e}\rangle = \sum_{\mu'\mu} e^*_{\mu'} e_\mu U_{\mu'\mu} \tag{1.21}$$

where e_μ are cyclic components of $| \vec{e}\rangle$ vector.

This operator allows to analyse the physical meaning of components of the matrix $U_{\mu'\mu}$. If $| \vec{e}\rangle$ is the same as the unit vector of the reference frame $| \vec{e}\rangle =| \mu\rangle$, the intensity will be proportional to the diagonal elements of the photon density matrix $I(\vec{e}_\mu) = U_{\mu\mu}$. Off-diagonal elements contribute to the intensity detected in other polarizations. For example, assuming

$$| \vec{e}\rangle =| \vec{e}_x\rangle = -2^{-1/2}(| 1\rangle- | -1\rangle)$$

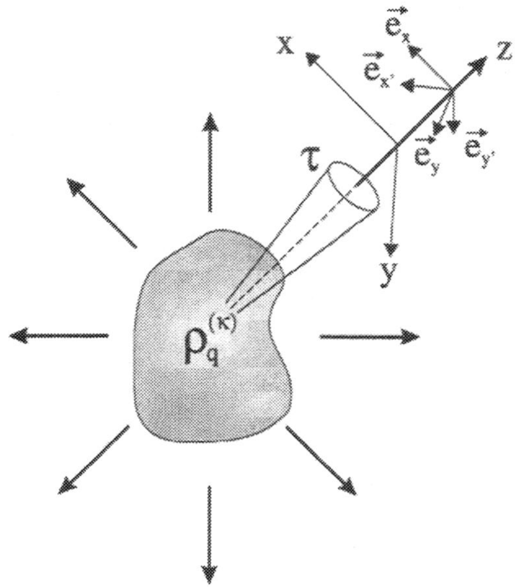

Figure 5. **Polarization, in the detector frame of reference, of a monochromatic light beam emitted by an atomic ensemble**

and

$$\mid \vec{e}\rangle = \mid \vec{e}_y\rangle = i2^{-1/2}(\mid 1\rangle + \mid -1\rangle)$$

we have

$$I(\vec{e}_x) = (U_{11} - U_{1-1} - U_{-11} + U_{-1-1})/2,$$

$$I(\vec{e}_y) = (U_{11} + U_{1-1} + U_{-11} + U_{-1-1})/2.$$

Polarization features of a light beam are described completely by the set of Stokes parameters, /13/ which in the detector frame of reference (x, y, z) are determined by (Fig. 5):

$$I = I_x + I_y = I_{45^o} + I_{-45^o} = I_+ + I_-$$

$$P = (I_x - I_y)/I; \qquad Q = (I_{45^o} - I_{-45^o})/I \qquad C = (I_+ - I_-)/I \quad (1.22)$$

where $I_{x,y,\pm 45^o}$ are the intensities of polarized beams with $\vec{e}_x$, $\vec{e}_y$, $\vec{e}_{x'} = 2^{-1/2}(\vec{e}_x + \vec{e}_y)$, $\vec{e}_{y'} = 2^{-1/2}(\vec{e}_x - \vec{e}_y)$ polarization vectors. I is the total intensity of a light beam, and P, Q, C are other Stokes parameters.

Making use of (1.21) and (1.22) one may express Stokes parameters in terms of the density matrix components /13/:

$$I = U_{11} + U_{-1-1} \qquad P = (U_{1-1} + U_{-11})/I$$

$$Q = -i(U_{1-1} - U_{-11})/I \qquad C = (U_{11} - U_{-1-1})/I \tag{1.23}$$

Photon density matrix in turn is expressed by Stokes parameters as follows:

$$U_{\mu\mu'} = \frac{I}{2}\begin{pmatrix} 1+C & -P+iQ \\ -P-iQ & 1-C \end{pmatrix}$$

In this manner the set of Stokes parameters, as well as the photon density matrix $U_{\mu\mu'}$, represent the complete description of the polarization characteristics of the light beam. Using these formulae in general one may obtain an expression for the Stokes parameters. But for practical purposes we shall use a more simple and straightforward method to obtain these expressions, following the results of the works /21, 23 / and (1.20), (1.21). Intensity of the light beam with polarization determined by $\vec{e}$ vector is given by:

$$I(\vec{e}) = B(\omega)N_a\tilde{\tau}\sum_{\mu'\mu} e^*_{\mu'}e_\mu \sum_{m_1mm'} \rho_{mm'}\langle Jm' \mid \hat{d}^+_\mu \mid J_1m_1\rangle\langle J_1m_1 \mid \hat{d}^+_{\mu'} \mid Jm\rangle$$

Introducing the scalar projection operator $P_{J_1} = \sum_{m_1} \mid J_1m_1\rangle\langle J_1m_1 \mid$, which is invariant with respect to rotations of the frame of reference, we have:

$$I(\vec{e}) = B(\omega)\tilde{\tau}N_a\sum_{\mu'\mu} e^*_{\mu'}e_\mu\langle \hat{d}^+_\mu \hat{P}_{J_1} d_{\mu'}\rangle_J$$

Using formulae of the spherical harmonic algebra we may write:

$$I(\vec{e}) = B(\omega)\tilde{\tau}N_a \mid \langle J_1 \,||\, \hat{d} \,||\, J\rangle \mid^2 \sum_{\mu\mu'} e^*_{\mu'}e_\mu(-1)^{J+J_1+\mu}$$

$$\sum_{\kappa q}\sqrt{2\kappa+1}\rho_q^{(\kappa)}\begin{Bmatrix} 1 & 1 & \kappa \\ J & J & J_1 \end{Bmatrix}\begin{pmatrix} 1 & 1 & \kappa \\ \mu' & -\mu & q \end{pmatrix} =$$

$$B(\omega)\tilde{\tau}N_a \mid \langle J_1 \,||\, \hat{d} \,||\, J\rangle \mid^2 \sum_{\kappa q}(-1)^{J+J_1}\sqrt{2\kappa+1}\rho_q^{(\kappa)}\begin{Bmatrix} 1 & 1 & \kappa \\ J & J & J_1 \end{Bmatrix}\Phi_q^{(\kappa)}(\vec{e}) \tag{1.24}$$

where

$$\Phi_q^{(\kappa)}(\vec{e}) = \sum_{\mu\mu'}(-1)^\mu\begin{pmatrix} 1 & 1 & \kappa \\ \mu' & -\mu & q \end{pmatrix} e^*_{\mu'}e_\mu$$

is the observation tensor introduced in /21/ which is transformed under the rotation of the reference frame in the same way as the irreducible tensor operator of the rank κ. Stokes parameters may be determined using the formulae (1.22) and (1.23) as

$$I = \frac{2B(\omega)N_a \sim \tau}{3(2J+1)^{1/2}} \mid \langle J \parallel \hat{d} \parallel J_1 \rangle \mid^2 I_0; \quad I_0 = \rho_0^{(0)} + \frac{2w(JJ_1)}{\sqrt{6}} \rho_0^{(2)}$$

$$P = w(JJ_1) I_0^{-1} (\rho_2^{(2)} + \rho_{-2}^{(2)}) \qquad Q = -iw(JJ_1) I_0^{-1} (\rho_2^{(2)} - \rho_{-2}^{(2)})$$

$$C = \frac{3}{\sqrt{2}} (2J+1)^{1/2} (-1)^{J+J_1} I_0^{-1} \left\{ \begin{array}{ccc} 1 & 1 & 1 \\ J & J & J_1 \end{array} \right\} \rho_0^{(1)} \qquad (1.25)$$

where

$$w(JJ_1) = \frac{3}{\sqrt{2}} (2J+1)^{1/2} (-1)^{J+J_1} \left\{ \begin{array}{ccc} 1 & 1 & 2 \\ J & J & J_1 \end{array} \right\}$$

These formulae show that the first rank polarization moment determine the circular polarization of optical radiation and the polarization moment of the second rank, the linear polarization. Classically, the multipole representation of the photon density matrix may be interpreted as the expansion of radiation of the excited atomic ensemble in terms of radiation of an isotropic oscillator - emitting nonpolarized light - a linear oscillator and a rotator.

1.2 Quadrupole ordering of angular momenta of excited atomic particles

Let us now turn to an examination of the physical reasons for the polarization of atomic states in ionized gas, restricting ourselves to a low temperature diluted plasma. The concentration of the excited and ionized particles in this case is much lower in comparison with the concentration of neutral particles. Optical emission in the linear spectrum results mainly from the direct excitation process of atoms by the electron impact or by self-absortion of resonance radiation in the plasma /24, 25/. Taking into consideration that the momentum of the inelasticaly scattered electron and the recoil momentum of an atomic particle are not accounted for in a plasma, the concept of axially symmetric system is used for the description of these processes /13, 23/. The velocity vector of the fast electron or direction of propagation of photons are taken for the natural symmetry axes for the elementary excitation processes.

1.2.1 EQUATION FOR POLARIZATION MOMENTS

Temporal variation of the statistic operator (1.2) describing the excited state with the energy E and angular momentum J, taking into account the radiative decay, relaxation, collisions and the radiation self-absorption, is given by the equation /17, 21/:

$$\frac{d\hat{\rho}_J}{dt} = -\hat{\hat{\Gamma}}\hat{\rho}_J + \hat{F}_J \qquad (1.26)$$

where $\hat{F}_j$ is the excitation operator and $\hat{\hat{\Gamma}}$ is the relaxation superoperator in the Liuville space R /13/. The physical meaning of the $\hat{F}$ operator is the rate of the statistical operator variation as a result of the direct excitation in a plasma with account for the formation of unequal populations of the sublevels of the excited state and the coherence, or phase correlation.

The main contribution to relaxation of the self-alignment of the excited neutral atoms in the low ionized gas is introduced by radiative decay and collisions with atoms of the gas and impurities in the ground state. Often in low temperature plasma these processes may be considered isotropic, because the neutral particles possess the isotropic Maxwellian velocity distribution. Mathematically, isotropy of the relaxation processes is represented by the invariance of the superoperator $\hat{\hat{\Gamma}}$ with respect to the rotation group. The action of $\hat{\hat{\Gamma}}$ superoperator on a certain tensorial operator $\hat{M}_q^{(\kappa)}$ in this case gives only the κ dependent factor Γ_κ :

$$\hat{\hat{\Gamma}}\hat{M}_q^{\kappa} = \Gamma_\kappa \hat{M}_q^{(\kappa)}$$

For isotropic conditions, the equation for the polarization moments (1.26) takes the form /21/:

$$\dot{\rho}_q^{(\kappa)} = -\Gamma_\kappa \rho_q^{(\kappa)} + F_q^{(\kappa)}$$

where Γ_κ is the relaxation constant of the polarization moment of the rank κ, and

$$F_q^{(\kappa)} = (2\kappa+1)^{1/2}\sum_{mm'}(-1)^{J-m'}\begin{pmatrix} J & J & \kappa \\ m' & -m & -q \end{pmatrix} F_{mm'} \qquad (1.27)$$

are the multipole moments of the relaxation operator:

$$\hat{F} = \sum_{\kappa q} F_q^{(\kappa)}\hat{T}_q^{(\kappa)}.$$

For stationary conditions ($\dot{\rho}_q^{(\kappa)} = O$), this equation takes the very simple form

$$\rho_q^{(\kappa)} = F_q^{(\kappa)}/\Gamma_\kappa$$

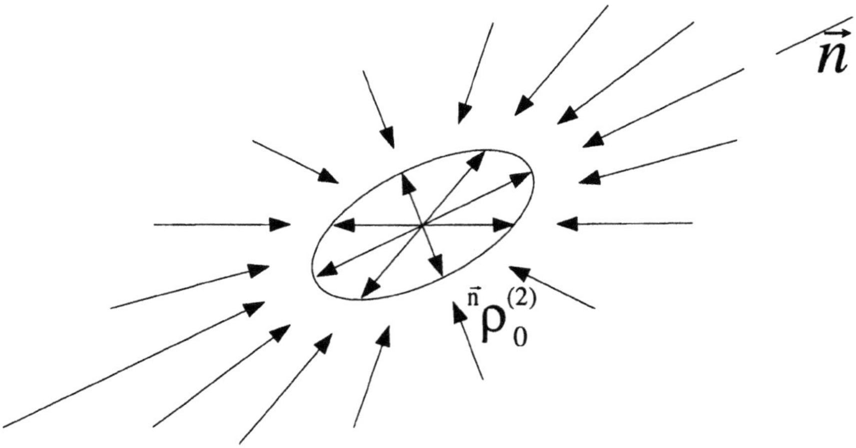

Figure 6. **Longitudinal alignment in the general case of axially symmetric excitation**

which may be explained thus: the type of ordering of momenta of the excited particles in ionized gas is determined by the multipole moments of the excitation tensor.

It should be noted that polarization of electron shells may also arise from the anisotropy of the relaxation process under collisions between atomic particles. Anisotropy of relaxation in this case is connected to the anisotropy in the space of the relative velocities of an emitting atomic particle and perturber. The drift of the heavy charged particles in the isotropic medium of neutral atoms may be the reason for anisotropy of relaxation in the discharge plasma. Mathematically the anisotropy of relaxation is reflected in the symmetry of $\hat{\hat{\Gamma}}$ superoperator under rotations of the frame of reference /17/.

Symmetry of electron impact excitation or radiation self-absorption in an elementary process is determined by the unit vector describing the exciting electron or the elementary light beam direction of propagation. It means that the orientation pseudovector ($\rho_q^{(1)}$) will not appear under the direct excitation in a plasma if the natural symmetry of excitation is not violated. Therefore the circular polarization of emission of the isotropic plasma without external field will not be observed /23/. Let us examine the case of the cylindrically symmetric plasma with respect to a certain axis $\vec{n}$ (Fig. 6). If the initial atomic state is spherically symmetric, the polarization moments with the nonzero quantum numbers q in the frame of reference with OZ axis oriented along $\vec{n}$ will not be induced because they are not invariant under rotations of the reference frame around $\vec{n}$:

$$^{\vec{n}}\rho_q^{(\kappa)} = {}^{\vec{n}}\rho_0^{(\kappa)}\delta_{0q}$$

where δ_{0q} is the Croniqer symbol. Transformation to laboratory frame of reference (x, y, z) is given by a Wigner matrix (1.5):

$$^{z}\rho_q^{(\kappa)}(\vec{n}) = D_{0q}^{(\kappa)^*}(\vec{n}z)\,{}^{\vec{n}}\rho_0^{(\kappa)} = C_q^{(\kappa)^*}(\vec{n})\,{}^{\vec{n}}\rho_0^{(\kappa)} \tag{1.28}$$

Stokes parameters of radiation of the axially symmetric plasma for observation at any direction determined by OZ axis of the detecter frame of reference may be expressed using (1.25) and (1.26) as follows:

$$\begin{aligned} P &= \left(\tfrac{3}{2}\right)^{3/2} \left\{ \begin{matrix} 1 & 1 & 2 \\ J & J & J_1 \end{matrix} \right\} {}^{\vec{n}}\rho_0^{(2)} \sin^2\theta \cos 2\varphi \left[\tfrac{(-1)^{J+J_1}}{\sqrt{2J+1}} \rho_0^{(0)} + \right. \\ &\left. \tfrac{3}{2\sqrt{6}} \left\{ \begin{matrix} 1 & 1 & 2 \\ J & J & J_1 \end{matrix} \right\} {}^{\vec{n}}\rho_0^{(2)} (3\cos^2\theta - 1) \right]^{-1} \\ Q &= \left(\tfrac{3}{2}\right)^{3/2} \left\{ \begin{matrix} 1 & 1 & 2 \\ J & J & J_1 \end{matrix} \right\} {}^{\vec{n}}\rho_0^{(2)} \sin^2\theta \sin 2\psi \left[\tfrac{(-1)^{J+J_1}}{\sqrt{2J+1}} \rho_0^{(0)} + \right. \\ &\left. \tfrac{3}{2\sqrt{6}} \left\{ \begin{matrix} 1 & 1 & 2 \\ J & J & J_1 \end{matrix} \right\} {}^{\vec{n}}\rho_0^{(2)} (3\cos^2\theta - 1) \right]^{-1} \end{aligned} \tag{1.29}$$

In the case of low degree of alignment $(\rho_q^{(2)} \ll \rho_0^{(0)})$ one may write:

$$P = P_0 \sin^2\theta \cos 2\varphi, \quad Q = P_0 \sin^2\theta \sin 2\varphi$$

where P_0 is the degree of polarization in the case when $\vec{n} \parallel OX$:

$$P_0 = (-1)^{J+J_1} \left(\frac{3}{2}\right)^{3/2} \sqrt{2J+1} \left\{ \begin{matrix} 1 & 1 & 2 \\ J & J & J_1 \end{matrix} \right\} {}^{\vec{n}}\rho_0^{(2)} / \rho_0^{(0)} \tag{1.30}$$

The values P_0 and $\rho_0^{(2)}$ characterize the longitudinal alignment of an excited state under the axially symmetric excitation. Classical analogy of such longitudinal alignment is the linear oscillator oriented along the alignment axis.

1.2.2 SELF-ALIGNMENT UNDER RADIATION SELF-ABSORPTION

Self-alignment under radiation self-absorption was examined in detail in the works/22, 26, 27/. Therefore we present here the general description of this mechanism, using the simpliest model of non-moving atoms where a macroscopic small atomic ensemble is irradiated by resonance light. Matrix elements of the statistic operator of the photon ensemble are represented by the formula

$$V_{\mu'\mu} = \overline{a_\mu a^*_{\mu'}}$$

where a_μ is the probability amplitude to find a photon in the state with the determined polarization.

Matrix elements of the excitation tensor will take the form:

$$\tilde{F}_{mm'} = B(\omega) \sum_{m_1 m' \mu\mu'} V_{\mu\mu'} \langle Jm \mid \hat{d}_\mu \mid J_1 m_1 \rangle \rho_{m_1 m'_1} \langle J_1 m'_1 \mid \hat{d}^+_{\mu'} \mid Jm \rangle \quad (1.31)$$

and the tensor of the optical excitation may be represented as:

$$\hat{F} = B(\omega) \sum_{\mu\mu'} V_{\mu\mu'} \hat{d}_\mu \hat{\rho}_{J_1} \hat{d}_{\mu'} \quad (1.32)$$

For ordinary low temperature plasmas, the degree of polarization of the ground state J_1 may be neglected compared to the angular anisotropy of the optical field resulting from the geometry and finite dimensions of the ionized gas entity. In this case

$$\rho_{m_1 m'_1} = N_1 \delta_{m_1 m'_1} / (2J_1 + 1) N_a,$$

where N_1 is the ground state density, N_a is the total density of the atomic particles. The multipole moment of the excitation tensor (1.27) may be expressed as:

$$\tilde{F}^{(\kappa)}_q = B(\omega) \frac{N_1 \sqrt{2\kappa + 1}}{N_a (2J_1 + 1)} \mid \langle J_1 \parallel \hat{d} \parallel J \rangle \mid^2 \left\{ \begin{array}{ccc} 1 & 1 & \kappa \\ J & J & J_1 \end{array} \right\}$$

$$\sum_{\mu\mu'} (-1)^{J + J_1 + \mu} V_{\mu\mu'} \left(\begin{array}{ccc} 1 & 1 & \kappa \\ \mu' & -\mu & q \end{array} \right)$$

or in the representation of polarization moments of the statistic photon operator (1.9), (1.17) we have:

$$\tilde{F}^{(\kappa)}_q = B(\omega) \frac{N_1 (-1)^{J + J_1}}{N_a (2J_1 + 1)} \mid \langle J_1 \parallel \hat{d} \parallel J \rangle \mid^2 \left\{ \begin{array}{ccc} 1 & 1 & \kappa \\ J & J & J_1 \end{array} \right\} V^{(\kappa)}_q \quad (1.33)$$

Hence the symmetry of the resonance optical field under self-absorption in a ionized gas determines the multipole moments of the excitation tensor of the same rank /22, 26/. Let $I(\vec{n}) d\vec{n}$ be the intensity of the elementary resonance light beam, propagating in an ionized gas along $\vec{n}$ within $d\vec{n}$ solid angle (Fig. 7). For an unpolarized elementary light beam, in the frame of

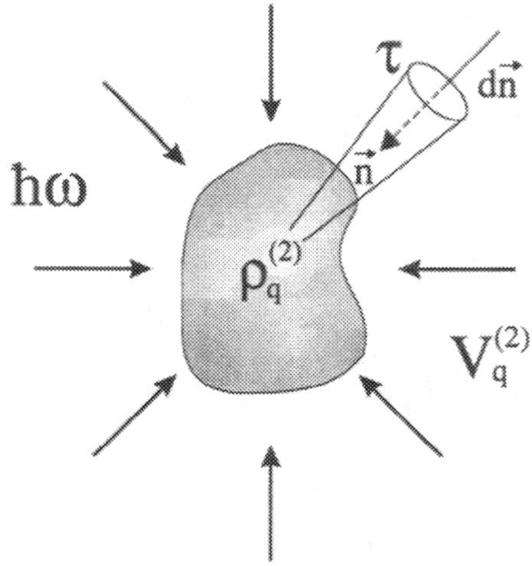

Figure 7. **Self-alignment of atoms under radiation self-absorption**

reference with a quantization axis directed along $\vec{n}$, the photon density matrix has the following components:

$$V_{11} = V_{-1-1} = 1/2,\ V_{00} = 0,\ V_{\mu\mu'} = 0,\ (\mu \neq \mu')$$

and according to (1.10) its polarization moments will take the form:

$$^{\vec{n}}V_q^{(\kappa)} = \frac{\sqrt{2\kappa+1}}{2}\left[\begin{pmatrix} 1 & 1 & \kappa \\ 1 & -1 & 0 \end{pmatrix} + \begin{pmatrix} 1 & 1 & \kappa \\ -1 & 1 & 0 \end{pmatrix}\right] I\delta_{0q} \qquad (1.34)$$

or

$$^{\vec{n}}V_0^{(0)} = I/\sqrt{3},\ ^{\vec{n}}V_q^{(2)} = I\delta_{0q}/\sqrt{6},\ ^{\vec{n}}V_q^{(\kappa)} = 0,\ \kappa = 1, \kappa \geq 3$$

and the non-zero components of the optical excitation tensor are expressed:

$$\tilde{F}_0 = \frac{(-1)^{J+J_1}B(\omega)N_1}{\sqrt{3}(2J_1+1)N_a}\left\{\begin{matrix} 1 & 1 & 0 \\ J & J & J_1 \end{matrix}\right\} \mid \langle J_1 \parallel \hat{d} \parallel J\rangle \mid^2; \quad \tilde{F}_1 = 0$$
$$\tilde{F}_2 = \frac{(-1)^{J+J_1}B(\omega)N_1}{\sqrt{6}(2J_1+1)N_a}\left\{\begin{matrix} 1 & 1 & 2 \\ J & J & J_1 \end{matrix}\right\} \mid \langle J_1 \parallel \hat{d} \parallel J\rangle \mid^2; \quad \tilde{F}_\kappa = 0, \kappa \geq 3 \qquad (1.35)$$

where I is the light intensity.

Transformation of the polarization moments to the laboratory frame of reference (x, y, z) is given by the following formula

$$^{z}V_q^{(\kappa)}(\vec{n}) = C_q^{(\kappa)^*}(\vec{n})\, ^{\vec{n}}V_0^{(\kappa)}.$$

The value of the multipole moments of the statistical photon operator for a certain part of ionized gas entity is calculated by integration over all directions $\vec{n}$:

$$^{z}V_q^{(\kappa)} = \int_{4\pi} d\vec{n}\, ^{z}V_q^{(\kappa)}(\vec{n}) = \int_{4\pi} d\vec{n} C_q^{(\kappa)*}(\vec{n})\, ^{\vec{n}}V_0^{(\kappa)} \tag{1.36}$$

Finally, making use of expressions (1.33) and (1.35), the polarization moments of the tensor of optical excitation of atomic ensemble in ionized gas are given by:

$$\tilde{F}_q^{(\kappa)} = B(\omega)\frac{N_1(-1)^{J+J_1}}{N_a(2J_1+1)} \mid \langle J_1 \parallel \hat{d} \parallel J\rangle \mid^2 \left\{ \begin{array}{ccc} 1 & 1 & \kappa \\ J & J & J_1 \end{array} \right\}$$

$$\int_{4\pi} d\vec{n} C_q^{(\kappa)*}(\vec{n})\, ^{\vec{n}}V_0^{(\kappa)} =$$

$$\tilde{F}_\kappa \int_{4\pi} d\vec{n} C_q^{(\kappa)*}(\vec{n}) I(\vec{n}) = \frac{4\pi}{2\kappa+1}\tilde{F}_\kappa I_q^{(\kappa)} \tag{1.37}$$

where

$$I_q^{(\kappa)} = \frac{2\kappa+1}{4\pi}\int_{4\pi} d\vec{n} C_q^{(\kappa)*}(\vec{n}) I(\vec{n})$$

is the multipole moment of intensity of radiation, or in other words the coefficients of the multipole representation of the angular dependence of intensity $I(\vec{n})$:

$$I(\vec{n}) = \sum_{\kappa=0}^{\infty}\sum_{q=-\kappa}^{\kappa} I_q^{(\kappa)} C_q^{(\kappa)}(\vec{n}). \tag{1.38}$$

Generalizing expression (1.37) for the case of optical excitation of many spectroscopic transitions i, we have:

$$\tilde{F}_q^{(\kappa)} = \frac{4\pi}{2\kappa+1}\sum_i \tilde{F}_\kappa(J_iJ) I_q^{(\kappa)}(J_iJ) \tag{1.39}$$

This simplified approach enables us to give a qualitative picture of the self-alignment distribution under radiation self-absorption in ionized gas entities of different geometrical shapes. For example, this technique was utilized in the studies of the spectropolarimetric effects in quiescent prominences of the solar atmosphere /11/.

1.2.3 SELF-ALIGNMENT BY ELECTRON IMPACT

Turning to the electron impact excitation process, we should emphasise that the quantitative methods of the spectropolarimetric diagnostic of ionized gases are based mainly on the collisional self-alignment process. The principal kinetic characteristic of electrons in ionized gas is the velocity distribution function $f(\vec{v})$ normalized to the total electron concentration N_e:

$$\int d\vec{v} f(\vec{v}) = N_e \tag{1.40}$$

The multipole representation in the velocity space or the expansion of $f(\vec{v})$ in terms of the orthonormal basis of the irreducible representations of the rotation group is implemented by the spherical harmonics $C_q^{(\kappa)}(\vec{n})$ ($\vec{n} = \vec{v}/v$, $v = |\vec{v}|$) as follows /28, 29/:

$$f(\vec{v}) = \sum_{\kappa=0}^{\infty} \sum_{q=-\kappa}^{\kappa} C_q^{(\kappa)}(\vec{n}) f_q^{(\kappa)}(v) \tag{1.41}$$

where

$$f_q^{(\kappa)} = \frac{2\kappa+1}{4\pi} \int_{4\pi} d\vec{n} C_q^{(\kappa)*}(\vec{n}) f(\vec{v})$$

The first term $f_0^{(0)}(v)$ describes the isotropic part of the velocity distribution function. It is connected to the electron concentration in an ionized gas:

$$\int_0^{\infty} f_0^{(0)}(v) v^2 dv = N_e / 4\pi \tag{1.42}$$

Electron motion anisotropy is reflected in the higher order harmonics. The dipole ordering of velocities or the local electron flux is characterized by the vector consisting of the cyclic components $f_q^{(1)}$. The tensor moment $f_q^{(2)}(v)$ is responsible for the quadrupole ordering of electron velocity vectors or the tensor of the density of the momentum flow η_{kl} /28/. Components of this tensor are proportional to the mean products of the Cartesian projections of the velocity vector:

$$\eta_{kl} = m\langle v_k v_l \rangle = m \int d\vec{v} v_k v_l f(\vec{v}) \tag{1.43}$$

where m is the electron mass.

In order to explain the electron impact self-alignment effect in ionized gas, let us examine the excitation of atoms by an elementary monokinetic electron beam with an electron density $f(\vec{v})d\vec{v}$ (Fig. 8). Excited atoms induced by collisions with the beam electrons will be axially symmetric with respect to rotation around $\vec{n}$. It means that in the frame of reference with the quantization axis aligned along $\vec{n}$, only longitudinal components of the excitation tensor will appear. For a monokinetic beam of unit intensity, these components are $\breve{F}_\kappa(v)$ and for a beam with density $f(\vec{v})d\vec{v}$ the excitation tensor will take the form ${}^z\breve{F}_0^{(\kappa)} = \breve{F}_\kappa(v) f(\vec{v}) d\vec{v}$. In the laboratory

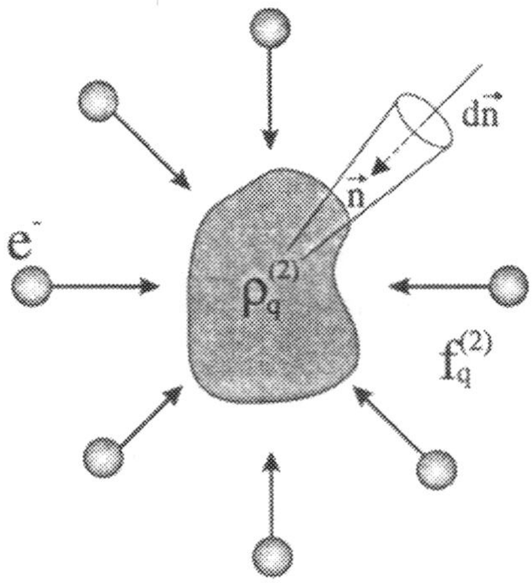

Figure 8. **Self-alignment of atoms under electron-atom collisions**

frame of reference (x, y, z) the components of the excitation tensor are:

$$^{z}\breve{F}_q^{(\kappa)}(\vec{n}) = C_q^{(\kappa)^*}(\vec{n})\breve{F}_\kappa(v)f(\vec{v})dv$$

For a macroscopic small volume of the ionized gas, the electron impact excitation tensor is expressed by integration in the velocity space:

$$^{z}\breve{F}_q^{(\kappa)} = \int d\vec{v}\, ^{z}\breve{F}_q^{(\kappa)}(\vec{v}) = \frac{4\pi}{2\kappa+1}\int_{v_t}^{\infty} \breve{F}_\kappa(v) f_q^{(\kappa)}(v)\, v^2\, dv \qquad (1.44)$$

As a result, the tensorial components of the electron impact excitation operator are determined by the multipole moments of the velocity distribution function of the same rank κ /12/.

1.2.4 CONNECTION OF STOKES PARAMETERS TO THE QUADRUPOLE MOMENT

Stokes parameters of the collisional emission of the macroscopic small volume of an ionized gas may be expressed in terms of the multipole moments of the velocity distribution function of fast electrons $f_q^{(\kappa)}$ as follows (1.25), (1.26), (1.44):

$$I = \frac{2}{3}(2\kappa+1)^{-1/2}B(\omega)N_a\tilde{\tau} \mid \langle J \parallel \hat{d} \parallel J_1\rangle \mid^2 I_0$$

$$P = \frac{4\pi w(JJ_1)}{5I_0\Gamma_2}\int_{v_t}^{\infty} \breve{F}_2(\nu) f_{2+}^{(2)}(v)\, v^2\, dv$$

$$Q = \frac{-i4\pi w(JJ_1)}{5I_0\Gamma_2} \int_{v_t}^{\infty} \breve{F}_2(\nu) f_{2-}^{(2)}(v)\, v^2\, dv$$

$$I_0 = \frac{4\pi}{\Gamma_0} \int_{v_t}^{\infty} \breve{F}_0(\nu) f_0^{(0)}(v)\, v^2\, dv + \frac{8\pi w(JJ_1)}{5\sqrt{6}\Gamma_2} \int_{v_t}^{\infty} \breve{F}_2(\nu) f_0^{(2)}(v)\, v^2\, dv \quad (1.45)$$

where $f_{2\pm}^{(2)}(\nu) = f_2^{(2)}(v) \pm f_{-2}^{(2)}(v)$ is calculated in the detector frame of reference.

If the ionized gas entity studied is axially symmetric with respect to a certain axis n, the only longitudinal components of the velocity distribution ${}^{\vec{n}}f_0^{(\kappa)}(v)$ will be in the frame of reference with OZ axis parallel to $\vec{n}$. For the Stokes parameters in this case we can write(1.29) :

$$\begin{aligned} P &= \left[R(JJ_1)^{-1} + \cos^2\theta - 1/3\right]^{-1} \sin^2\theta \cos 2\varphi \\ Q &= \left[R(JJ_1)^{-1} + \cos^2\theta - 1/3\right]^{-1} \sin^2\theta \sin 2\varphi \end{aligned} \quad (1.46)$$

$$R(JJ_1) = \frac{\frac{1}{5}\sqrt{\frac{3}{2}} w(JJ_1) \frac{\Gamma_0}{\Gamma_2} \int_{v_t}^{\infty} \breve{F}_2(v)\, {}^{\vec{n}}f_0^{(2)}(v)\, v^2\, dv}{\int_{v_t}^{\infty} \breve{F}_0(v) f_0^{(0)}(v)\, v^2\, dv} \quad (1.47)$$

For the low degree of alignment $\rho_0^{(2)} \ll \rho_0^{(0}$ which takes place, for instance, in the low temperature laboratory discharges, the function $R(JJ_1)$ represents the degree of polarization of atomic ensemble P in the frame of reference with $OX \parallel \vec{n}$ (1.30).

In order to analyse the physical meaning of the kernels of these integral equations (1.45) we point out that the longitudial components of the excitation tensor of atomic ensemble excited by an electron beam of unit density, $\breve{F}_\kappa(v)$ are connected to the components of the excitation tensor $\breve{F}_{mm'}$ in the representation $J - m$(1.27):

$$\breve{F}_\kappa(v) = (2\kappa + 1)^{1/2} \sum_m (-1)^{J-m} \begin{pmatrix} J & J & \kappa \\ m & -m & 0 \end{pmatrix} \breve{F}_{mm}(\nu) \quad (1.48)$$

The last term in turn describes the rate of the change of the diagonal elements of the density matrix $\breve{F}_{mm} = \dot{\rho}_{mm}$ or the population of the magnetic sublevels of the studied state under electron impact. This rate may be expressed in terms of the cross-section of excitation $\sigma(v)$: $\breve{F}_{mm} = \sigma_m(v)v$. Consequently, we may write: $\breve{F}_\kappa(v) = v \cdot \sigma^{(\kappa)}(v)$, where

$$\sigma^{(\kappa)}(v) = (2\kappa + 1)^{1/2} \sum_m (-1)^{J-m} \begin{pmatrix} J & J & \kappa \\ m & -m & 0 \end{pmatrix} \sigma_m(v) \quad (1.49)$$

is the cross-section of excitation of the polarization moment of the rank κ. For $\kappa = 0$:

$$\sigma^{(0)}(\nu) = (2J+1)^{1/2} \sum_m \sigma_m(v) = (2J+1)^{1/2} \sigma(v)$$

The cross-section of the excitation of the zero rank is proportional to the total cross-section of the level $\sigma(v)$. The cross-section of the electron impact alignment takes the form:

$$\sigma^{(2)}(v) = \sqrt{5}\,[(2J+3)(2J+1)(J+1)J(2J-1)]^{-1/2} \sum_m (3m^2 - J(J+1))\sigma_m(v) \tag{1.50}$$

Finally we have:

$$P = \frac{8\pi w(JJ_1)}{5m^2\Gamma_2 I_0} \int_{\varepsilon_t}^{\infty} \sigma^{(2)}(\varepsilon) f_{2+}^{(2)}(\varepsilon)\varepsilon d\varepsilon$$

$$Q = \frac{-i8\pi w(JJ_1)}{5m^2 I_0 \Gamma_2} \int_{\varepsilon_t}^{\infty} \sigma^{(2)}(\varepsilon) f_{2-}^{(2)}(\varepsilon)\varepsilon d\varepsilon$$

$$R(JJ_1) = \frac{\frac{1}{5}\sqrt{\frac{3}{2}}\, w(JJ_1)\frac{\Gamma_0}{\Gamma_2} \int_{\varepsilon_t}^{\infty} \sigma^{(2)}(\varepsilon)\, \vec{n} f_0^{(2)}(\varepsilon)\varepsilon d\varepsilon}{\int_{\varepsilon_t}^{\infty} \sigma^{(0)}(\varepsilon) f_0^{(0)}(\varepsilon)\varepsilon d\varepsilon} \tag{1.51}$$

where $\varepsilon_t = mv^2/2$ is the threshold electron energy for the excitation of the level J ,

$$I_0 = \frac{8\pi}{m^2\Gamma_0} \int_{\varepsilon_t}^{\infty} \sigma^{(0)}(\varepsilon) f_0^{(0)}(\varepsilon)\varepsilon d\varepsilon + \frac{16\pi w(JJ_1)}{5\sqrt{6}m^2\Gamma_2} \int_{\varepsilon_t}^{\infty} \sigma^{(2)}(\varepsilon) f_0^{(2)}(\varepsilon)\varepsilon d\varepsilon$$

By making use of these expressions one may analyse new possibilities of spectropolarimetric diagnostics. Measuring Stokes parameters of linear impact radiation of an ionized gas on different spectral transitions makes it possible to determine the quadrupole moment of the velocity distribution function of fast electrons, which brings principally new information about an ionized gas entity. Quadrupole moment of the velocity distribution function cannot be measured by the traditional methods of spectroscopy intensity diagnostic. Study of the distribution of the spectropolarimetric characteristics enables one to estimate the energy exchange in the different parts of a ionized gas and to examine the relative role of those parts in the energy balance of an entity. By these means the spectropolarimetry allows the remote sensing of energetic structural characteristics of an ionized gas entity. Such possibilities are of principal importance for the quantitative study of ionized gases with a high degree of anisotropy (solar atmosphere, ionosphere boundary regions of plasmas, beams of charged particles).

1.3 Electron impact alignment cross sections

A principal question connected with spectropolarimetric diagnostic is the determination of the electron impact alignment cross-section $\sigma^{(2)}(\varepsilon)$, which mathematically represents the kernel of the integral equations of polarization spectroscopy (1.51). In this paragraph we shall discuss the ways to obtain information about $\sigma^{(2)}(\varepsilon)$ using theoretical methods and data of spectrocopic and collisional experiments.

1.3.1 RELATIONS TO COLLISIONAL SPECTROSCOPIC PARAMETERS

The cross-section of electron impact alignment by its physical definition represents the collisional quadrupole ordering of angular momenta and may be expressed as the combination of the cross sections of excitation of magnetic sublevels σ_m (1.50). In contrast to the excitation cross-section, the alignment cross-section has different signs, depending on the type of the angular distribution of momenta (Fig. 9). Alignment cross-section is zero if the cross-sections of excitation of magnetic sublevels are equal or the total cross-section is sphericaly symmetric. The limit case of the anisotropic excitation corresponds to the population of the only one magnetic sublevel ($m = 0$):

$$\frac{\sigma^{(2)}}{\sigma} = -\sqrt{5}\left[\frac{J(J+1)}{(2J+3)(2J+1)(2J-1)}\right]^{1/2} < 0$$

or the sublevels $m = \pm J$:

$$\frac{\sigma^{(2)}}{\sigma} = \sqrt{5}\left[\frac{J(2J-1)}{(2J+3)(2J+1)(J+1)}\right]^{1/2} > 0$$

For the others cases of the population distribution the ratio $\sigma^{(2)}/\sigma$ lies between these extreme values. The most simple experimental method for determining electron impact cross-section is to measure the degree of polarization P_0 of emission of atoms, under stationary electron beam excitation, for observation at a right angle. In this case, in the frame of reference with the quantization axis along the beam of electrons, the excitation tensor has only longitudinal components:

$$^{n}\breve{F}_0^{(\kappa)} = f\breve{F}_\kappa(v) = f\sigma^{(\kappa)}(v)v$$

where f is the electron concentration in the beam.

Polarization moments will take the form $^{n}\rho_0^{(\kappa)} = fv\sigma^{(\kappa)}(v)/\Gamma_\kappa$ and the polarization of the optical emission will be given by:

$$P_0(v) = \left[R^{-1}(v) - 1/3\right]^{-1} \tag{1.52}$$

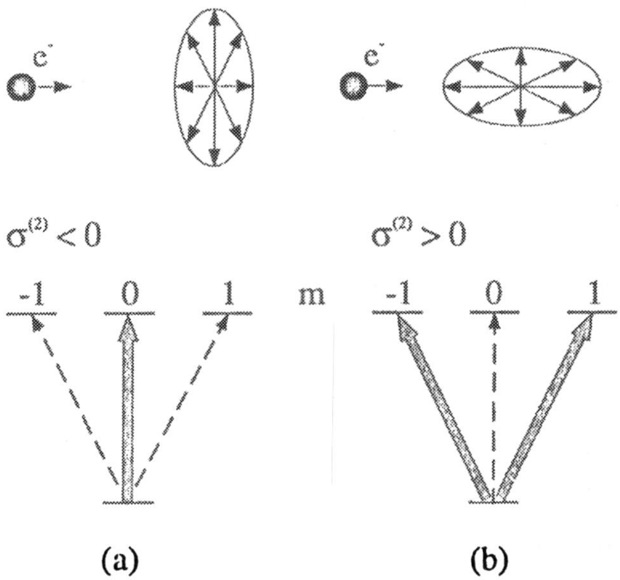

Figure 9. **Impact alignement excitation cross-section at threshold (a) and high energies (b)**

where

$$R(v) = \left(\frac{3}{2}\right)^{3/2} (-1)^{J+J_1}(2J+1) \left\{ \begin{array}{ccc} 1 & 1 & 2 \\ J & J & J_1 \end{array} \right\}^{-1} \frac{\Gamma_0 \sigma^{(2)}(v)}{\Gamma_2 \sigma(v)}$$

Therefore the alignment cross section may be expressed as:

$$\sigma^{(2)}(v) = \left(\frac{3}{2}\right)^{3/2} \frac{\Gamma_2(-1)^{J+J_1}}{\Gamma_0(2J+1)} \left\{ \begin{array}{ccc} 1 & 1 & 2 \\ J & J & J_1 \end{array} \right\}^{-1} \frac{P_0(v)\sigma(v)}{1+P_0(v)/3} \qquad (1.53)$$

Direct reliable $P_0(v)$ measurements have been implemented only for a small number of cases involving the relatively strong spectral transitions in helium /30/. Measurements, especially in the threshold region of energies of the exciting electrons, are very complicated because of the low intensity of the optical emission. In view of this, R.H. McFarland /31/ suggested measuring the radiation intensities $I(90^o)$, $I(\beta)$ at the angles 90° and β with respect to the direction of the beam. The dependence $P_0(v)$ in this case takes the form:

$$P_0(v) = \left[\frac{I(\beta, v)}{I(90^o, v)} - 1\right] \cos^{-2}\beta$$

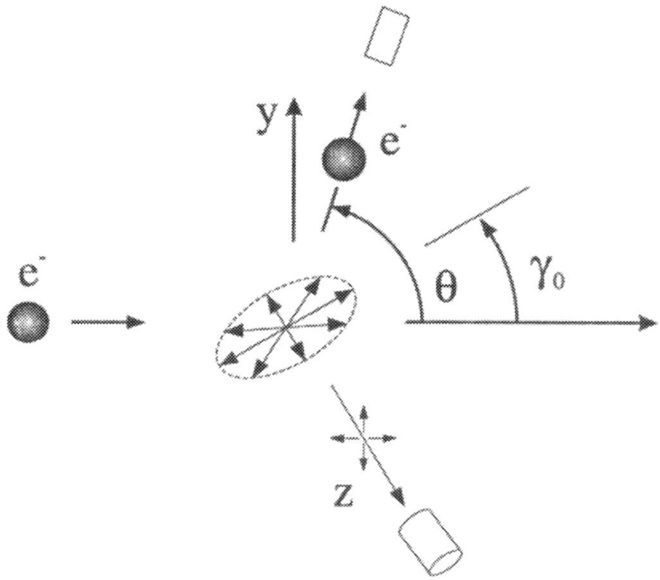

Figure 10. **Scheme of the experimental measurement of electron-photon angular correlations by observation at right angle to the scattering plane**

and the electron impact alignment cross section is given by:

$$\sigma^{(2)}(v) = 3\left(\frac{2}{3}\right)^{3/2} \frac{\Gamma_2}{\Gamma_0} \frac{(1-)^{J+J_1}}{2J+1} \left\{ \begin{array}{ccc} 1 & 1 & 2 \\ J & J & J_1 \end{array} \right\} \sigma(v)$$
$$\left[I(\beta, v)/I(90^o, v) - 1\right] \left[2 - 3\sin^2\beta + I(\beta, v)/I(90^o, v)\right]^{-1} \tag{1.54}$$

The other possible method for the determining $\sigma^{(2)}(v)$ is the use of the coherent effects under pulsed or modulated impact excitation, measuring the dependence of the modulation of the linearly polarized radiation as a function of the energy of the exciting electrons /32/.

For a wider group of states the cross-section of the electron impact alignment may be obtained making use of the bank of experimental and theoretical data of the angular correlation parameters /33/. In such experiments the coincidence between the inelastically scattered electrons and the polarization or angular characteristics of emitted photons is studied. For experiments of this type, the axial symmetry of the system with respect to the beam direction is violated. The polarization of the atomic ensemble is therefore not described only in terms of the longitudinal components of the polarization moments of atomic density matrix. Polarization moments of odd ranks and with q not equal to zero also exist. A frame of reference with

OZ axis directed orthogonally to the scattering plane is used for describing the polarization effects in the experiments with axial symmetry violation (Fig. 10). In this case, besides Stokes parameters P, Q, C, the following set of parameters is introduced: $P_1 = (P^2 + Q^2)^{1/2}$: maximal polarization degree which is attained at the angle of analyser $\gamma_0 = (1/2)\arctan(Q/P)$, the mean z-projection of the total angular momentum of electron shell $\langle J_\perp \rangle$ which is proportional to the Stokes parameter C. For describing the experiments on electron-photon angular correlations, Fano parameters are also used /13, 23/:

$$O_{1-} = \frac{\langle J_z \rangle}{J(J+1)}; \quad A_{1+} = \frac{\langle J_y J_x + J_x J_y \rangle}{J(J+1)}$$
$$A_0 = \frac{\langle 3J_z^2 - J^2 \rangle}{J(J+1)}; \quad A_{2+} = \frac{\langle J_y^2 - J_z^2 \rangle}{J(J+1)} \tag{1.55}$$

Polarization of the states with the total angular moment $J = 1$ is described completely by the following parameters /33/:

$$\lambda = \sigma_0/\sigma = | a_0 |^2 / (| a_0 |^2 + 2 | a_1 |^2)$$

and $\chi = \arg(a_1/a_0)$, where σ_0 is the cross-section and of $m = 0$ substate excitation , $\sigma = \sigma_0 + \sigma_1 + \sigma_{-1}$ is the total excitation cross-section, and $a_{0,\pm 1}$ are the complex amplitudes of the electron impact excitation of the separate sublevels.

Table 1 gives formulae connecting different parameters used for describing the spectropolarimetric effects in electron-photon angular correlation experiments. Using these relationships and the definition of $\sigma^{(2)}$ (1.50), one obtain expressions for the electron impact alignment cross-section in terms of the angular correlations parameters.

For $J = 1$ we have:

$$\begin{aligned} \sigma^{(2)}(v) = &\ \tfrac{2\pi}{\sqrt{6}} \int_0^\pi d\theta \, \sin\theta \, \sigma(\theta, v) \, [1 - 3\lambda(\theta, v)] \\ \sigma^{(2)}(v) = &\ \tfrac{\pi}{\sqrt{6}} \int_0^\pi d\theta \, \sin\theta \, \sigma(\theta, v) \, [1 + 3P(\theta, v)] \\ \sigma^{(2)}(v) = &\ \tfrac{\pi}{\sqrt{6}} \int_0^\pi d\theta \, \sin\theta \, \sigma(\theta, v) \, [1 + 3P_l(\theta, v) \cos 2\gamma_0(\theta, v)] \end{aligned} \tag{1.56}$$

For $J \neq 1$ one may write:

$$\sigma^{(2)}(v) = 2\pi\sqrt{5} \left[\frac{J(J+1)}{(2J+3)(2J+1)(2J-1)} \right]^{1/2} \int_0^\pi d\theta \, \sin\theta \, \sigma(\theta, v) A_0(\theta, v) \tag{1.57}$$

where $\sigma(\theta, v)$ is the differential scattering cross section.

TABLE 1. Relationship between different spectropolarimetric parameters used in electron-photon angular correlation experiments: $(P, Q, C), (P_e, \gamma_0, \langle J_\perp \rangle)$ - sets of parameters in the detector frame of reference with OZ axis ortogonal to the scattering plane - and $(\lambda, \chi), (A_q, O_q)$ - sets of parameters for a $0-1$ spectral transition within the collision frame of reference

	P Q C	$P_e\, \gamma\, \langle I_\perp \rangle$	$\lambda\ \chi$	$A_0 A_{1+} A_{2+} O_{1-}$
P		$P_e cos2\gamma_0$	$2\lambda - 1$	$-(4A_0 + 1)/3$
Q		$P_e sin2\gamma_0$	$-2\sqrt{\lambda(1-\lambda)}cos\chi$	$-2A_{1+}$
C		$-\langle J_\perp \rangle$	$2\sqrt{\lambda(1-\lambda)}sin\chi$	$-2O_{1-}$
P_e^2	$P^2 + Q^2$		$(2\lambda - 1)^2$ $+4\lambda(1-\lambda)cos^2\chi$	$\frac{1}{9}(4A_0 + 1)^2 + 4A_{1+}$
$tg2\gamma_0$	Q/P		$-2\sqrt{\lambda(1-\lambda)}$ $/(2\lambda - 1)$	$6A_{1+}/(4A_0 + 1)$
$\langle J_\perp \rangle$	$-C$		$-2\sqrt{\lambda(1-\lambda)}sin\chi$	$2O_{1-}$
λ	$(P+1)/2$	$\frac{1}{2}(P_e cos2\gamma_0 + 1)$		$-(2A_0 - 1)/3$
$tg\chi$	C/Q	$-\langle J_\perp \rangle / P_e sin2\gamma_0$		O_{1-}/A_{1+}
A_0	$-(1+3P)/4$	$-\frac{1}{4}(1 + P_e cos2\gamma_0$	$(1-3\lambda)/2$	
A_{1+}	$-Q/2$	$-\frac{1}{2} sin2\gamma_0$	$\sqrt{\lambda(1-\lambda)}cos\chi$	
A_{2+}	$(P-1)/4$	$(P_e cos\gamma_0 - 1)/4$	$(\lambda - 1)/2$	
O_{1-}	$-C/2$	$\langle J_\perp \rangle / J(J+1)$	$-\sqrt{\lambda(1-\lambda)}sin\chi$	

1.3.2 METHODS OF CALCULATION

For high energies of exciting electrons the impact alignment cross section may be obtained in Born approximation. This problem was examined in /34/, using symmetry considerations for the electron impact excitation operator. Amplitudes of the Zeeman sublevels of atomic states C_{J_m} are proportional to the matrix elements of the electron impact excitation operator:

$$\sum_e \exp\{-i\vec{k}'\hat{\vec{r}}_e\}$$

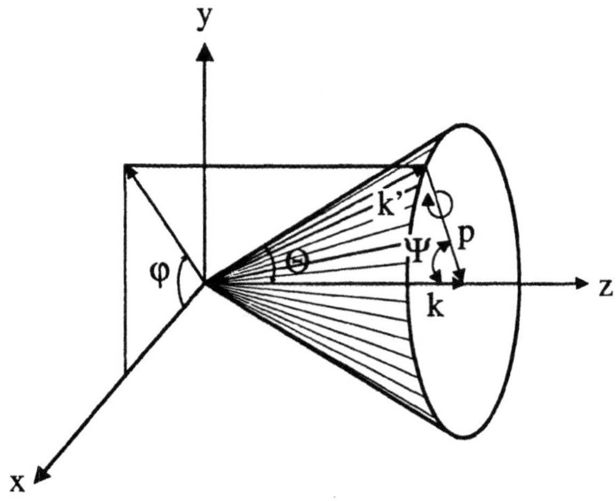

Figure 11. **Calculation of Born electron impact alignment cross section ($\vec{k}$ and $\vec{k}'$ are the wave vectors of incident and scattered electrons)**

where $\hat{\vec{r}}_e$ are atomic electron coordinates, $\vec{k}, \vec{k}_1$ are wave vectors of the incident and scattered electrons, $\vec{k}' = \vec{k} - \vec{k}_1$, (Fig. 11).

This operator is invariant with respect to rotation around the axis parallel to $\vec{k}'$ vector. Therefore, if the ground state has zero orbital and spin moments, in the frame of reference with the quantization axis along $\vec{k}'$ only the elements having a projection of momentum on this axis that is null will be excited /13, 23/. The spin projection on the quantization axis may only change its value $\Delta S = \pm 1$ resulting from the exchange interaction at the intercombination transitions. Hence under the impact excitation by the fast electrons in the frame of reference with OZ axis parallel to $\vec{k}'$ only the substates with $m = O$ for the spin conservation collisions ($\Delta S = O$) may be excited and for intercombination transitions ($\Delta S = \pm 1$) the states with $m = \pm 1$ projections.

Within this frame of reference, the atomic ensemble studied may be characterized by the noncoherent superposition of states with different m values; it is therefore possible to describe this situation in terms of the cross section of excitation $\sigma(\vec{k}')$. Taking into consideration the angular dependence of the wave vector $\vec{k}' = \vec{k}'(\theta, \varphi)$, $\sigma(\vec{k}')$ is the inelastic differential cross-section. The cross-sections of higher orders, and in particular the alignment cross-section, must be also considered.

For LS coupling approximation, the states of the electron shell may be described by the spin quantum number. For the alignment cross-section, in this approximation, we have using (1.50):

$$\sigma^{(2)}(\vec{k}') = \alpha(S)\sigma(\vec{k}') \tag{1.58}$$

where

$$\alpha(S) = \sqrt{5}\left[\frac{J(J+1)}{(2J+3)(2J+1)(2J-1)}\right]^{1/2}\left[\frac{3S}{J(J+1)} - 1\right].$$

Transition to the collision frame of reference with the quantization axis oriented along $\vec{k}$ (laboratory frame of reference) is realized by the rotation matrix. After this transformation the transversal components of the polarization moments with $q \neq O$ will also appear. These components are dependent on the asymuthal angle φ as exp(iqφ) and will vanish after integration within the range $[0, 2\pi]$. The total alignment cross section which is obtained as a result of the averaging over all contributions from various directions will have only the longitudinal component:

$$\sigma^{(2)}(v) = \int_0^{2\pi} d\varphi \int_0^{\pi} d\theta \sin\theta\, \sigma^{(2)}(\vec{k}')\left[\frac{3}{2}\cos\Psi(\theta) - \frac{1}{2}\right]$$

where $\Psi(\theta)$ is the angle between $\vec{k}$ and $\vec{k}'$ vectors (Fig. 11) expressed as:

$$\cos^2\Psi(\theta) = \frac{1 + \delta\cos^2\theta - 2\sqrt{\delta}\cos\theta}{1 + \delta - 2\sqrt{\delta}\cos\theta}$$

$\delta = k_1^2/k^2 = \varepsilon_1/\varepsilon$, ε is the energy of the incident electron, $\varepsilon_1 = \varepsilon - \varepsilon_t$ energy of the inelasticaly scattered electron, ε_t the threshold energy of the level studied.

As a result, for the alignment cross-section under these approximations we have:

$$\sigma^{(2)}(v) = 3\pi\alpha(S)\int_0^{\pi} d\theta \sin\theta\sigma(\theta, v)\frac{1 + \delta\cos^2\theta - 2\sqrt{\delta}\cos\theta}{1 + \delta - 2\sqrt{\delta}\cos\theta} \tag{1.59}$$

By these means of calculating the Born alignment cross section, it is possible to use the data base of experimental and theoretical inelastic differential cross sections $\sigma(\theta, v)$.

In order to estimate the alignment cross section at threshold energies it is easy to consider this question on the basis of the momentum conservation law. This approach was used in the earlier works on electron impact excitation and developed in a number of later studies /13, 23/. For even isotopes with zero orbital and spin momenta in the ground state, let the energy of the exciting electron moving along OX axis of the laboratory frame be equal to the threshold excitation energy of the level under study. Neglecting relativistic effects, the projection of the orbital momentum of the incident electron with respect to OX axis is null. The projection of the spin moment on the same axis is $\pm 1/2$. According to Percival-Seaton /13/, spin-spin and spin-orbit interactions may be neglected under collisions. In other words, no electron has its spin changed during collision. The scattered electron has zero kinetic energy and momentum. Therefore the $0X$ projection of orbital momentum is invariant under the threshold collisions. Spin variation may occur only for an electron exchange process. Hence for the threshold value of the alignment cross-section we may write

$$\sigma_t^2 = \alpha(S)\sigma_t \tag{1.60}$$

where σ_t is the threshold electron impact excitation cross-section.

Using more sophisticated models of electron-atom systems, calculations of the spectroscopic collisional parameters for the threshold excitation are presented in many works. We describe here the basic approach using the simple two-level scheme and non-relativistic approximation /34/, neglecting spin-orbit and exchange interactions. Two channels of the collisional process may be regarded: elastic scattering with the conservation of energy and atomic impact excitation. Let collisions induce the transition E_0, $L = O \rightarrow E_1$, L (L is the orbital moment).

The total Hamiltonian of the system incident electron-atom (H) consists of the atomic Hamiltonian H_0 , the Hamiltonian of the free electron and electron-atom interaction operator V_{int}. The temporal dependence of the wave function of the system is described

$$\psi = \psi(\vec{x}) \exp\left[-i(\varepsilon + E_0)t/\hbar\right]$$

(ε is the energy of the incident electron), and the coordinate part $\psi(\vec{x})$ ($\vec{x}$ are coordinates of all electrons) is given by the equation:

$$(\Delta + k^2 - \hat{H}_a - \hat{V}_a)\psi(\vec{x}) = 0 \tag{1.61}$$

where $k = \sqrt{2m\varepsilon}/\hbar$ is the wave number of the incident electron $\hat{H}_a = 2m\hat{H}_0/\hbar^2$, $\hat{V}_a = 2mV_{int}/\hbar^2$ and Δ is the Laplasian of the free electron.

In the collision frame of reference we have:

$$\hat{H}_a\Phi(\vec{x}) = E_0\Phi(\vec{x}) \;\; \hat{H}_a\varphi_M(\vec{x}) = E_1\varphi_M(\vec{x}) \tag{1.62}$$

where $\Phi(\vec{x})$ and $\varphi_M(\vec{x})$ are atomic wave functions in the ground and excited states and M indicates magnetic substates of the upper level.

The normalization and orthogonality relationships:

$$\int d\vec{x} \mid \Phi(\vec{x}) \mid^2 = 1 \quad \int d\vec{x}\, \Phi^*(\vec{x})\varphi_M(\vec{x}) = 0$$

$$\int d\vec{x}\, \varphi^*_M(\vec{x})\varphi_{M'}(\vec{x}) = \delta_{MM'}$$

enable the expansion of the wave function $\psi(\vec{x})$ of the electron-atom system in terms of Φ and φ_M functions:

$$\psi(\vec{x}) = \tilde{F}(\vec{r})\Phi(\vec{x}) + \sum_M \tilde{f}_M(\vec{r})\varphi_M(\vec{x}) \tag{1.63}$$

where $\tilde{F}(\vec{r})$ and $\tilde{f}_M(\vec{r})$ are functions of the coordinate $\vec{r}$ of the exciting electron. Initial conditions are determined by the asymptotic behavior of the wave function of the incident electron out of the zone of the electron-atom interaction. The asymptotic wave function consists of the incident plane wave of the free electron and of the spherical scattered wave:

$$\tilde{F}(\vec{r}) = \exp\{ik\vec{e}_z\vec{r}\} + A(\vec{n})\frac{\exp\{ikr\}}{r}; \quad \tilde{f}_M(\vec{r}) = a_M(\vec{n})\frac{\exp\{ik_1r\}}{r}$$

For $r > R_0$ (R_0 is the radius of the interaction sphere),

$$k_1 = [2m(\varepsilon + E_0 - E_1)]^{1/2}/\hbar$$

is the wave number of the inelastically scattered electron, $A(\vec{n}), a_M(\vec{n})$ are amplitudes of elastic and inelastic scattering, $\vec{n} = \vec{r}/r$.

The functions $\tilde{F}(\vec{r})$ and $\tilde{f}_M(\vec{r})$ for $r > R_0$ are the solutions of equation of motion of the free electron:

$$(k^2 + \Delta)\tilde{F}(\vec{r}) = 0 \quad (k_1^2 + \Delta)\tilde{f}_M(\vec{r}) = 0$$

$A(\vec{n})$ and $a_M(\vec{n})$ are connected to the differential cross-sections of elastic $\sigma_e(\vec{n})$ and inelastic $\sigma_M(\vec{n})$ scattering:

$$\sigma_e(\vec{n}) = \mid A(\vec{n}) \mid^2 \quad \sigma_M(\vec{n}) = (k_1/k) \mid a_M(\vec{n}) \mid^2$$

The total cross-section of excitation of the sublevel M of the excited state will be given by:

$$\sigma_M = \frac{k_1}{k}\int d\vec{n} \mid a_M(\vec{n}) \mid^2 \tag{1.64}$$

Equation (1.61) is usually solved expressing the wave function of outer electrons in terms of an expansion in spherical harmonics, which is known as partial wave representation. This expansion takes the form /34/:

$$\tilde{F}(\vec{r}) = \frac{1}{r}\sum_{\Lambda=0}^{\infty} C_0^{(\Lambda)}(\vec{n})\tilde{F}^{(\Lambda)}(r); \; \tilde{f}_M(\vec{r}) = \frac{1}{r}\sum_{\Lambda=0}^{\infty} C_{-M}^{(\Lambda)}(\vec{n})\tilde{f}_{-M}^{(\Lambda)}(r) \quad (1.65)$$

where the coefficients are expressed:

$$\begin{aligned} \tilde{F}^{(\Lambda)}(r) &= \tfrac{4\pi}{2\Lambda+1} r \int d\vec{n}\, C_0^{(\Lambda)^*}(\vec{n})\tilde{F}(\vec{r}) \\ \tilde{f}_{-M}^{(\Lambda)}(r) &= \tfrac{4\pi}{2\Lambda+1} r \int d\vec{n}\, C_{-M}^{(\Lambda)^*}(\vec{n})\tilde{f}_{-M}(\vec{r}) \end{aligned}$$

substituting the asymptotic expansion and using the relation /16/:

$$\exp\{ik\vec{r}\cdot\vec{e}_z\} = \sum_{\Lambda=0}^{\infty}(2\Lambda+1)i^{\Lambda}j_{\Lambda}^{(}kr)C_0^{(\Lambda)}(\vec{n})$$

we have that for $r > R_0$:

$$\begin{aligned} \tilde{F}^{(\Lambda)}(r) &= (2\Lambda+1)i^{(\Lambda)}rj_{\Lambda}(kr) + \exp[i(kr - \pi\Lambda/2)]A^{(\Lambda)} \\ \tilde{f}_{-M}^{(\Lambda)}(r) &= \exp[i(k_ir - \pi\Lambda/2)]a_M^{(\Lambda)} \end{aligned}$$

where $j_{\Lambda}(k,r)$ are Bessel spherical functions and $A^{(\Lambda)}$ and $a_M^{(\Lambda)}$ the partial scattering amplitudes.

The connection to the scattering amplitudes is given by formulae:

$$A^{(\Lambda)} = e^{i\Lambda\pi/2}\tfrac{4\pi}{2\Lambda+1}\int d\vec{n}\, C_0^{(\Lambda)^*}(\vec{n})A(\vec{n})$$

$$A(\vec{n}) = \textstyle\sum_{\Lambda=0}^{\infty} e^{-i\Lambda\pi/2}C_0^{(\Lambda)}(\vec{n})A^{(\Lambda)}$$

$$a_{-M}^{(\Lambda)} = e^{i\Lambda\pi/2}\tfrac{4\pi}{2\Lambda+1}\int d\vec{n}\, C_{-M}^{(\Lambda)^*}(\vec{n})a_M(\vec{n})$$

$$a_{-M}(\vec{n}) = \textstyle\sum_{\Lambda=0}^{\infty} e^{-i\Lambda\pi/2}C_{-M}^{(\Lambda)}(\vec{n})a_M^{(\Lambda)}$$

Hence, using (1.65) we have:

$$\sigma_M = \frac{k_1}{k}\sum_{\Lambda=0}^{\infty}\frac{2\Lambda+1}{4\pi} \mid a_{-M}^{(\Lambda)} \mid^2 = \sum_{\Lambda=0}^{\infty}\sigma_{M,\Lambda}, \quad (1.66)$$

where

$$\sigma_{M,\Lambda} = \frac{k_1}{k}\frac{2\Lambda+1}{4\pi} \mid a_{-M}^{(\Lambda)} \mid^2$$

are the partial cross-sections of excitation of a sublevel $\varphi_M(\vec{x})$.

The partial wave-expansion technique enables us to reduce the problem of the electron-atom scattering to a system of connected unidimensional differential equations. Making use of straightforward transformations, we have:

$$\begin{aligned}
&\sum_\Lambda \Phi(\vec{x})C_0^{(\Lambda)}(\vec{n})\left[k^2+\frac{\partial^2}{\partial r^2}-\frac{\Lambda(\Lambda+1)}{r^2}\right]\tilde{F}^{(\Lambda)}(r)\\
&+\sum_{\Lambda M}\varphi_M(\vec{x})C_{-M}^{(\Lambda)}(\vec{n})\left[k_1^2+\frac{\partial^2}{\partial r^2}-\frac{\Lambda(\Lambda+1)}{r^2}\right]\tilde{f}_{-M}^{(\Lambda)}(r)\\
=\;&\hat{V}_a(\vec{x},\vec{r})\left[\sum_\Lambda \Phi(\vec{x})C_0^{(\Lambda)}(\vec{n})\tilde{F}^{(\Lambda)}(r)+\sum_{\Lambda M}\varphi_M(\vec{x})C_{-M}^{(\Lambda)}(\vec{n})\tilde{f}_{-M}^{(\Lambda)}(r)\right]
\end{aligned}$$

Using the orthogonality relations of atomic wave functions we have the following system:

$$\begin{aligned}
\left(\frac{\partial^2}{\partial r^2}+k^2+\frac{\Lambda(\Lambda+1)}{r^2}\right)\tilde{F}^{(\Lambda)}(r)=\;&\sum_{\Lambda'}\langle\Lambda\mid\hat{V}_a\mid\Lambda'\rangle\tilde{F}^{(\Lambda')}(r)\\
&+\sum_{M'\Lambda'}\langle\Lambda\mid\hat{V}_a\mid M'\Lambda'\rangle\tilde{f}_{-M'}^{(\Lambda')}(r)\\
\left(\frac{\partial^2}{\partial r^2}+k^2+\frac{\Lambda(\Lambda+1)}{r^2}\right)\tilde{f}_{-M}^{(\Lambda)}(r)=\;&\sum_{\Lambda'}\langle M\Lambda\mid\hat{V}_a\mid\Lambda'\rangle\tilde{F}^{(\Lambda')}(r)\\
&+\sum_{M'\Lambda'}\langle M\Lambda\mid\hat{V}_a\mid M'\Lambda'\rangle\tilde{f}_{-M'}^{(\Lambda')}(r)
\end{aligned}\tag{1.67}$$

where

$$\begin{aligned}
\langle\Lambda\mid\hat{V}_a\mid\Lambda'\rangle &= \int d\vec{x}\,d\vec{n}\,\Phi^*(\vec{x})C_0^{(\Lambda)^*}(\vec{n})\hat{V}_a\Phi(\vec{x})C_0^{(\Lambda')}(\vec{n})\\
\langle\Lambda\mid\hat{V}_a\mid M'\Lambda'\rangle &= \int d\vec{x}\,d\vec{n}\,\Phi^*(\vec{x})C_0^{(\Lambda)^*}(\vec{n})\hat{V}_a\varphi_{M'}(\vec{x})C_{-M}^{(\Lambda')}(\vec{n})\\
\langle M\Lambda\mid\hat{V}_a\mid\Lambda'\rangle &= \int d\vec{x}\,d\vec{n}\,\varphi_M^*(\vec{x})C_{-M}^{(\Lambda)^*}(\vec{n})\hat{V}_a\Phi(\vec{x})C_0^{(\Lambda')}(\vec{n})\\
\langle M\Lambda\mid\hat{V}_a\mid M'\Lambda'\rangle &= \int d\vec{x}\,d\vec{n}\,\varphi_M^*(\vec{x})C_{-M}^{(\Lambda)^*}(\vec{n})\hat{V}_a\varphi_{M'}(\vec{x})C_{-M}^{(\Lambda')}(\vec{n})
\end{aligned}$$

These equations, together with the boundary conditions, determine the partial amplitudes and according to (1.66) the cross-sections of the magnetic sublevel excitation.

One of the most effective methods of calculating the polarization parameters, especially in the threshold energy region, is the R-matrix technique /35/, which enables us to describe the resonance structure of the $\sigma^{(2)}(\varepsilon)$ dependence in more detail. The multiresonance structure of electron impact cross-section has been experimentally observed /23/. The analysis of the reliability of the methods of determination of the electron impact alignment cross-section was performed by means of comparative calculations and the results were presented in /36/.

CHAPTER 2

EXPERIMENTAL METHODS OF POLARIMETRIC SPECTRAL SENSING

Turning to experimental ionized gas spectropolarimetry, we should point out that this technique can be usefully applied to the following principal problems: measurement of the Stokes parameters of an atomic line spectrum and derivation from observed spectropolarimetric data of the local spectropolarimetric parameters in the radiation emitting layers.

2.1 Optical polarimetry techniques

Polarimetry is the investigation of the polarization properties of radiation of ionized gases. Despite a long history of studies, a substantial improvement of the polarimetry technique is still possible. The polarization of radiation can be very complicated and usually only incomplete information is available. Since we are normally interested in radiation with a narrow spectral interval, we have to deal with low intensity signals for which photon statistics become significant and measurements in many cases are performed near the sensitivity threshold. At the same time, even for relatively small fluctuations of the radiation intensity, temporal and spatial fluctuations of the polarization signal may become significant, which gives rise to complications of the measurement procedure. Spectropolarimetric studies and plasma spectroscopy in general are based on the solution of ill-posed problems. It means that small uncertainties in the measured quantities may correspond to considerable uncertainties in the local plasma parameters. The further requirement of highly precise measurements has, therefore, to be satisfied.

2.1.1 REQUIREMENTS FOR EXPERIMENTAL POLARIMETRIC SCHEMES

In the following, we now formulate the conditions that the equipment used to investigate the polarization of atomic ensembles, by the measure of the polarization of plasma radiation /37/, must satisfy:

-possibility of simultaneous measurement of all the Stokes parameters of the fluctuating radiation,

-absence of moving parts in order to achieve a rapid response, and a high precision of measurement,

-loss of light as low as possible,

-possibility of combination with a high-resolution spectrometer.

Advances in the techniques used to measure the polarization of plasma radiation proceeded for a long time separately from the substantial advances in ellipsometry. Many different polarimeters for plasma studies were developed between the late 1920s and 1970s, but they were all based on the comparison of light intensities with particular orthogonal plane polarizations. The final result was calculated from formulae such as

$$P = \left(\frac{I_{\|}}{I_{\perp}}\frac{I_{\perp}^{c}}{I_{\|}^{c}} - 1\right)\left(\frac{I_{\|}}{I_{\perp}}\frac{I_{\perp}^{c}}{I_{\|}^{c}} + 1\right)^{-1}$$

where the superscript c labels calibration measurements. The quantities $I_{\|}$ and $I_{\perp}$ could be measured simultaneously so that the experimental arrangement usually employed the splitting of the light flux associated with a particular spectral line into two beams, using polarizing prisms or crystal plates. The radiation passing through the analyzer was recorded simultaneously by two photodetectors (usually photomultipliers), and the signals were each fed into a difference-circuit. In another method, the intensities of the vibrations with perpendicular polarizations were measured in succession, by rotating the analyzer. It is readily seen that even when instrumental uncertainties are reduced, these methods are suitable only for measuring high degrees of polarization. However, the uncertainty of the calculated polarization P usually exceeds by a substantial factor the uncertainty of the measured radiation intensity.

The common feature of the above methods, which is a universal source of uncertainty in the measured degree of polarization, is the differing transmission of the spectral device for perpendicularly polarized beams of radiation. This uncertainty is usually removed by depolarizing the radiation incident on the entrance slit of the spectrometer. This is done by depolarizing prisms, or suitably oriented quarter-wave plates.

In addition, each scheme has its own particular sources of uncertainty. In the arrangement using beam splitting, uncertainties are due to the nonuniformity of the plasma radiation over the beam cross section as well as differences between optical channels, including differences between the sensitivities of the photodetectors. The last of these can be eliminated by suitable calibration. The arrangement in which the beams of light are not recorded simultaneously is preferred in investigations of stationary objects. The specific uncertainties introduced in this method are associated with the nonuniform rotation of the analyzer and beats of its axis. Moreover, when the plasma parameters are subject to strong fluctuations and drift, so that the radiation is similarly affected, this method requires a long averaging time or gives only qualitative results.

2.1.2 FOURIER POLARIMETER

A polarization spectrometer for the investigation of low radiation intensities from non-homogeneous plasmas combines the advantages of the above two methods. The two beams of light with orthogonal polarizations are recorded by channel-switching counting circuits, so that rotating elements are avoided.

The polarimeter described by Kazantsev et al./37/ employs the simultaneous detection of the radiation fluxes and performs absolute measurements of the degree of polarization for non-inhomogeneous nonstationary objects with a low level of optical signal. The two orthogonal plane polarizations are separated by a Rochon prism, which produces a beam divergence angle of 5.7^o.

In addition to the deficiencies of the above methods that we have already mentioned (low precision, low sensitivity, and slow response), they suffer from one further disadvantage. In principle, they are capable of measuring only two Stokes parameters that correspond to plane polarized radiation, and even then they require *a priori* information on the position of the axes of the polarization ellipse. It follows that the quality of the final results always depends on the validity of the adopted assumptions. Moreover, the range of modern techniques used in complete polarimeters, i.e., polarimeters that simultaneously measure all the Stokes parameters, is very restricted.

A new Fourier polarimeter for plasma investigations which satisfies the above criteria was therefore developed. A block diagram of this polarimeter is shown in Fig. 12 and we shall now describe it in detail. The radiation to be analyzed is collimated and then intercepted by a linear electro-optic modulator (EM) supplied by an alternating voltage from the oscillator OSC. The incident wavefront is then divided in two by the analyzer consisting of the two polarizing prisms $P1$, $P2$ mounted so that the optical axes of one them ($P1$) coincide with the optical axes of the modulator ($\alpha = O$), whereas the axes of the other are at an angle of 45° to them. Since the polarizing prisms divide the incident radiation into two beams with perpendicular polarizations, the analyzer has four optical outputs. Radiation from these outputs is first wavelength-selected and is then intercepted by photodetectors $PH1 - PH4$.

We shall use the Muller matrices to describe the operation of the system. At any of the four analyzer outputs the Stokes vector is a function of the incident-wave Stokes vector $[\eta]$:

$$[\eta'] = [M_A][D(\alpha)][M_\delta][\eta]$$

where $[M_A]$, $[M_\delta]$ are the Muller matrices of the analyzer and phase modulator, respectively, and $[D(\alpha)]$ is the rotation matrix. Substituting the cor-

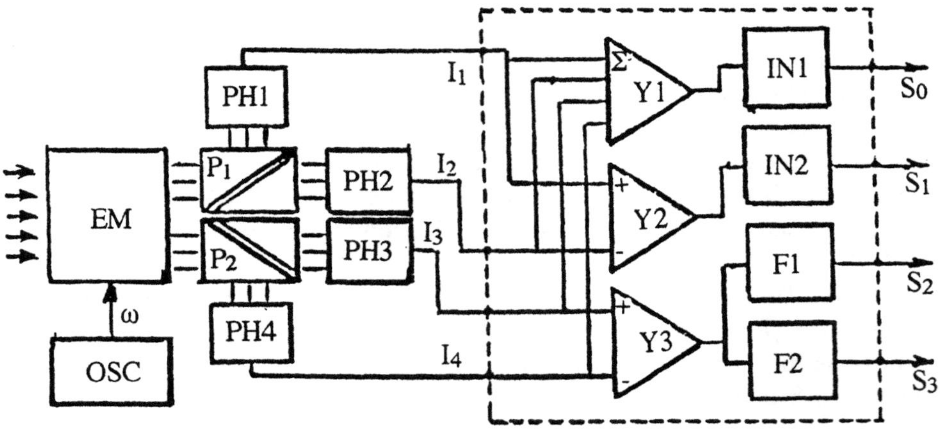

Figure 12. Scheme of a Fourier spectropolarimeter

responding matrices and solving for the intensity of light at the analyzer outputs, we obtain

$$2\eta_0' = \eta_0 + \eta_1 \cos 2\alpha + (\eta_2 \cos\delta + \eta_3 \sin\delta)\sin 2\alpha.$$

Let us now suppose that the phase angle δ varies periodically with frequency ω. Expanding $\sin\delta$ and $\cos\delta$ into a Fourier serie, and neglecting the higher-order harmonics, at the four photodetector outputs we obtain respectively the following signals:

$$\begin{aligned} I_1(\alpha = 0^o) &= \tfrac{1}{4}(\eta_0 + \eta_1), \\ I_2(\alpha = 90^o) &= \tfrac{1}{4}(\eta_0 - \eta_1), \\ I_3(\alpha = 45^o) &= \tfrac{1}{4}[\eta_0 + \eta_2(2\tilde{J}_2 \cos 2\omega t + \tilde{J}_0) + 2\eta_3\tilde{J}_1 \sin\omega t], \\ I_4(\alpha = 135^o) &= \tfrac{1}{4}[\eta_0 + \eta_2(2\tilde{J}_2 \cos 2\omega t + \tilde{J}_0) - 2\eta_3\tilde{J}_1 \sin\omega t]; \end{aligned}$$

where $\tilde{J}_m$ are Bessel functions. It is clear from these equations that only two of the four polarimeter channels need to be used to measure the Stokes parameters uniquely. The parameters η_0 and η_1 are determined from the constant component of the photodetector current, η_2 is determined from the component at frequency 2ω, and η_3 from the component at frequency ω. The use of four channels removes the dependence of the results on small misalignments, and increases the sensitivity of the instrument.

Since the optical channels are identical, their transmission characteristics are also identical. The polarimeter can therefore be combined with a spectrometer that has several optical channels and is placed between the polarizing prisms and the photodetectors. The polarization transmission characteristic of the spectrometer must then be taken into account in the relevant equations. The speed of response of the instrument is then restricted by the frequency characteristic of the electro-optic modulator and by the response time of the signal detection system. In principle, a response time of a few tens of nanoseconds can be achieved.

Because of the need for high experimental precision, for statistical analysis of several runs, and real-time processing of the data, the entire system has to be run by a computer. An apparatus for the investigation of polarization spectra, based on these principles, is described in /37/. It incorporates a somewhat simplified version of the Fourier polarimeter described above, and is capable of measuring the parameters η_0, η_1 and η_2. The electro-optic modulator is a DKDP crystal activated by a high-voltage acoustic oscillator. The analyzer consists of two polarizing prisms with optic axes at 45° to one another. Spectral analysis is performed by a DFC-24 double spectrometer, which simultaneously analyzes the two optical channels. The decoupling between the channels is $-50dB$. Two optical-fiber lightguides are mounted on the input slit of the spectrometer and are coupled to cooled photodetectors. Synchronous detection at combination frequencies is used to isolate the useful signal.

The outputs of the synchronous detectors are fed into a CAMAC system working in conjunction with a microcomputer. The system records the polarization spectra of individual spectral lines and of segments of the continuum in their neighbourhood, and performs a statistical analysis of the data.

2.1.3 POLARIMETERS FOR SOLAR SPECTRA STUDIES

In this field, polarimetric measurements have a long history and have been used in different problems of solar physics, mostly for Zeeman magnetometry. However, the use of spectropolarimetry as a diagnostic method for the local sensing of the solar atmosphere is also connected to the measurement of the degree of impact polarization in the center of emission lines. Most observations of this type have been made in the hydrogen emission line during chromospheric solar flares.

The first ground based bservations were made at Paris Observatory /38/ with a filter tuned at the center of H_α line. The choice of this line made it possible to use a flare patrol instrument and to benefit of the support available for flare patrol. Moreover, even if H_α may not be ultimately the best line to study the impact plarization, its large turbulent width strongly reduces the net polarization by the Zeeman and Stark effects. The instrument used was an H_α flare patrol heliograph using a 15 cm refractor telescope. The axial symmetry of the optics insured that those measurements would not be affected by instrumental polarization. The monochromatic Lyot filter had a 0.75 Åbandpass. A rotating half-wave plane introduces a modulation of the intensity $\mathcal{I}$ of the radiation that has travelled through the linear polarizer which follows it:

$$\mathcal{I} = \frac{1}{2}(I - P\cos 4\alpha - Q\sin 4\alpha),$$

where Q and P are the linear Stokes parameters and α is the angle between the direction of the transmission of the polarizer and the axis of the half-wave plate. The Stokes parameter C is not measured. The two entrance and exit sides of the half-wave plate are not strictly parallel. Therefore, the wave-plate rotation introduces a motion of the image. Consequently, a cross correlation technique had to be used to compensate for these motions. Cross-correlation was made using digitized data obtained on area not affected by the flares. Then, by substraction of the appropriate digitized images, the Stokes parameters were obtained. In order to reduce the effect of the temporal variation of the flare intensity I, a set of three consecutive images was used. This gives:

$$P_{2i+1} = (-1)^i \left[(\mathcal{I}_{2i} + \mathcal{I}_{2i+2}) - 2\mathcal{I}_{2i+1}\right]$$

$$Q_{2i} = (-1)^i \left[(\mathcal{I}_{2i-1} + \mathcal{I}_{2i+1}) - 2\mathcal{I}_{2i}\right],$$

with

$$\alpha = \frac{\pi}{8} \times i,$$

where i varies from one to seven. The polarization fraction and the azimuth of the polarization are easily derived from the Stokes parameters. The same type of spectropolarimetric measurements have been also realized on the large solar telescope of Baykal Astrophysical Observatory in Irkutsk /39/ with the help of an analyser of linear polarization mounted behind the exit slit of the spectrograph. This analyser was made of two crossed Wollaston prisms fixed one after the other. Ordinary and extraordinary beams were propagating parallel to each other behind this system and the distance between the beams was varied, changing the distance between the prisms. Intensities of the ordinary and extraordinary rays were equalized by rotating the achromatic $\lambda/2$ plate mounted behind the Wollaston prism. Another $\lambda/2$ plate was mounted before the slit in order to determine the azimuth of the polarization plane. Every studied location was exposed three times on one photoplate, for three positions of the phase $\lambda/2$ plate corresponding to 0, 22.5 and 45 degree. Linear polarization parallel to the grooves of the diffraction grating of the spectrometer (first spectral band) was extracted at the first position of the $\lambda/2$ plate. When $\lambda/2$ plate was fixed at 45 degrees, the polarization plane was turned at 90 degrees compared to the zero orientation of the phase plate. Therefore, the third spectral band corresponded to a polarization parallel to the direction of dispersion of the grating. At a 22.5 degree angle the $\lambda/2$ plate gave two spectral bands with analysis of the linear polarizations oriented at 45 degrees with respect to the initial position.

2.2 Magnetic spectropolarimetry

In the study of laboratory gas discharges, the aims of the magnetic spectropolarimetry is to measure the dependence of the polarization of optical emission as a function of the weak magnetic field applied to the part of the discharge under study. Magnetic field destruction of the coherence of the states gives rise to a characteristic magnetic field dependence of polarization, known as Hanle effect /3/. The profile of the Hanle effect signals or self-alignment signal is very informative. It reflects the degree of self-alignment, orientation of the local alignment axes in the detector frame of reference and relaxation constants of the excited state. The important advantage of the magnetic method is the high degree of sensitivity which is very valuable for measurements in laboratory discharges with a low degree of atomic self-alignment. Besides, this method is free from systematic errors introduced by the polarization of the optical elements of experimental set-up and walls of the discharge tube or the instrumental polarization.

2.2.1 POLARIZATION MOMENTS IN PRESENCE OF A MAGNETIC FIELD. HANLE EFFECT

The time evolution equation for statistical operator $\hat{\rho}_J$ taking into account the weak magnetic field, is given by /13, 21/:

$$\rho_J = \frac{1}{i\hbar}[\hat{V}_M \rho_J] - \hat{\hat{\Gamma}}\hat{\rho}_J + \hat{F} \tag{2.1}$$

where

$$[\hat{V}_M \rho_J] = \hat{V}_M \rho_J - \rho_J \hat{V}_M, \ \hat{V}_M = \mu_0 g_J \vec{H}\hat{\vec{J}}$$

is the Hamiltonian of the interaction of the electron shell with the outer magnetic field, μ_0 is Bohr's magneton, g_J is Lande g-factor of the studied level, and $\vec{H}$ is magnetic field.

A weak magnetic field used for realization of the magnetic polarization technique does not significantly influence the kinetic of particles in plasma and the redistribution of light beams; therefore, the excitation tensor $\hat{F}$ and the relaxation superoperator may be regarded as magnetic field independent functions.

We then introduce the Larmor precession frequency operator $\vec{\Omega} = \mu_0 g_J \vec{H}/\hbar$ of the electron shell in external magnetic field and make some transformations, taking into account the isotropy of the relaxation superoperator $\hat{\hat{\Gamma}}$:

$$\dot{\rho}_q^{(\kappa)} = \frac{\vec{\Omega}}{i} Sp\left(\hat{T}_q^{(\kappa)}[\hat{\vec{J}}\hat{\rho}_J]\right) - \Gamma_\kappa \rho_q^{(\kappa)} + F_q^{(\kappa)} \tag{2.2}$$

Using (1.4) and expanding $\vec{\Omega}$ and $\hat{\vec{J}}$ vectors in terms of cyclic coordinates in the laboratory frame of reference for the first term of the right hand side we have:

$$\frac{\Omega}{i} Sp\left(\hat{T}_q^{(\kappa)}[\hat{\vec{J}}\hat{\rho}_J]\right) = (i)^{-1} \sum_{\mu\kappa' q} \Omega_\mu \rho_q^{(\kappa)} (-1)^q \sum_{mm'm''} \langle m \mid \hat{T}_{-q}^{(\kappa)} \mid m' \rangle \cdot$$
$$\langle m' \mid J_\mu \mid m'' \rangle \langle m'' \mid T_{q'}^{(\kappa')} \mid m \rangle - \langle m \mid \hat{T}_{-q}^{(\kappa)} \mid m' \rangle \langle m' \mid \hat{T}_{q'}^{(\kappa)'} \mid m'' \rangle \langle m'' \mid J_\mu \mid m' \rangle$$

which is transformed using the Wigner-Eckert theorem, the expression for reduced matrix elements

$$\langle J \parallel \hat{T}^{(\kappa)} \parallel J \rangle = \sqrt{2\kappa+1},\ \langle J \parallel \hat{J} \parallel J \rangle = \sqrt{J(J+1)(2J+1)},$$

formulae from the theory of angular momentum and symmetry conditions for 3 and 6j symbols /13/ to the form:

$$\frac{\Omega}{i} Sp\left(\hat{T}_q^{(\kappa)+}[\hat{\vec{J}}\hat{\rho}_J]\right) = (i)^{-1} \mu\kappa' q' \sum \left[(2\kappa+1)(2\kappa'+1)J(J+1)(2J+1)\right]^{1/2}$$
$$(-1)^{q+2J+\kappa+\kappa'+1} \Omega_\mu \rho_{q'}^{(\kappa')} \begin{Bmatrix} \kappa & \kappa' & 1 \\ J & J & J \end{Bmatrix} \begin{pmatrix} \kappa & 1 & \kappa' \\ -q & \mu & q' \end{pmatrix} \left[1 - (-1)^{\kappa+\kappa'}\right] \tag{2.3}$$

Analysing this result we should note that $\kappa = \kappa'$ and $q' = q - \mu$. Then the evolution equation may be written in the following form:

$$\begin{aligned} \dot{\rho}_q^{(\kappa)} &= -i(-1)^{\kappa+q} \sqrt{\kappa(\kappa+1)(2\kappa+1)} \sum_{\mu=0,\pm1} \Omega_\mu \rho_{q-\mu}^{(\kappa)} \begin{pmatrix} \kappa & 1 & \kappa \\ -q & \mu & q-\mu \end{pmatrix} \\ &\quad -\Gamma_\kappa \rho_q^{(\kappa)} + F_q^{(\kappa)} \end{aligned}$$

or finally:

$$\begin{aligned} \dot{\rho}_q^{(\kappa)} &= -iq\Omega_0 \rho_q^{(\kappa)} - \tfrac{i}{\sqrt{2}} \sqrt{(\kappa-q+1)(\kappa+q)}\Omega_1 \rho_{q-1}^{(\kappa)} \\ &\quad + \tfrac{i}{\sqrt{2}} \sqrt{(\kappa+q+1)(\kappa-q)}\Omega_{-1} \rho_{q+1}^{(x)} + 1 - \Gamma_\kappa \rho_q^{(\kappa)} + F_q^{(\kappa)} \end{aligned} \tag{2.4}$$

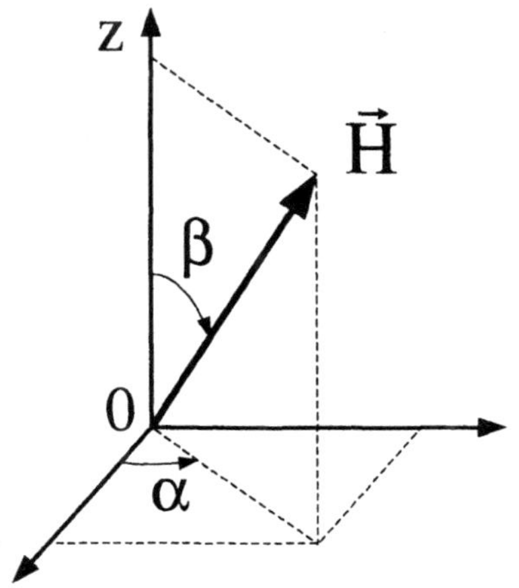

Figure 13. **Transfer from laboratory frame of reference (XYZ) to the frame of reference with OZ' axis parallel to the magnetic field**

For the most simple case of a magnetic field oriented parallely to the OZ axis of the laboratory frame of reference, $\Omega_0 = |\vec{\Omega}|$, $\Omega_{\pm 1} = 0$. Therefore, we have:

$$\dot{\rho}_q^{(\kappa)} = -iq\Omega\rho_q^{(\kappa)} - \Gamma_\kappa\rho_q^{(\kappa)} + F_q^{(\kappa)} \tag{2.5}$$

The expression (2.4) for an arbitrarily oriented magnetic field ($q = 0$, ± 1, ± 2) is transformed into a system of differential equations, the solution of which is,in general, a rather complicated problem. As a rule it is easier to make a transformation to the frame of reference with the quantization axis oriented along $\vec{H}$, in which the polarization moments ${}^{\vec{H}}\rho_q^{(\kappa)}$ are given by (1.5) and then to go back to the initial frame of reference (Fig. 13):

$$
\begin{aligned}
{}^{\vec{H}}\rho_q'^{(x)} &= \sum_q D_{qq'}^{(\kappa)*}(z,\vec{H})\,{}^{z}\rho_q^{(\kappa)} = \sum_q D_{qq'}^{(\kappa)*}(z,\vec{H})\,{}^{z}F_q^{(\kappa)}/(\Gamma_\kappa + iq'\Omega) \\
{}^{z}\rho_q^{(\kappa)} &= \sum_{q'} D_{q'q}^{(\kappa)*}(\vec{H},z)\,{}^{\vec{H}}\rho_{q'}^{(\kappa)} = \sum_{q'q''} D_{qq'}^{(\kappa)}(z\vec{H})D_{q''q'}^{(\kappa)}(z\vec{H})\,{}^{z}F_{q''}^{(\kappa)}/(\Gamma_\kappa + iq'\Omega)
\end{aligned}
\tag{2.6}
$$

Using these equations for the polarization moments one may express the Stokes parameters of the studied radiation and magnetic field dependence or the shape of the self-alignment signal.

2.2.2 STOKES PARAMETERS DETERMINATION

For the conventional laboratory experimental scheme, the magnetic field is directed along the line of sight. If the degree of alignment of atomic particles is rather low ($\rho_0^{(2)} \ll \rho_0^{(0)}$) the degree of polarization is given:

$$P(H) = w(JJ_1)\frac{\rho_2^{(2)} + \rho_{-2}^{(2)}}{\rho_0^{(0)}} \tag{2.7}$$

where

$$w(JJ_1) = \frac{3}{2}(-1)^{J+J_1}\sqrt{2J+1}\begin{Bmatrix} 1 & 1 & 2 \\ J & J & J_1 \end{Bmatrix}$$

Under stationary conditions, ($\dot{\rho}_q^{(\kappa)} = 0$), the magnetic field dependence of the polarization moments takes the form

$$\rho_q^{(\kappa)} = F_q^{(\kappa)}/(\Gamma_\kappa + iq\Omega)$$

and the expression for the polarization degree may be written as:

$$P(H) = \frac{w(JJ_1)}{\rho_0^{(0)}}\left(\Gamma_2\frac{F_2^{(2)} + F_{-2}^{(2)}}{\Gamma_2^2 + 4\Omega^2} + i\Omega\frac{F_2^{(2)} - F_{-2}^{(2)}}{\Gamma_2^2 + 4\Omega^2}\right) \tag{2.8}$$

where $F_q^{(\kappa)} = \Gamma_\kappa \rho_q^{(\kappa)}(0)$, ($\rho_q^{(\kappa)}(0)$ is the polarization moment for zero magnetic field).

The degree of polarization then takes the form (2.7):

$$P(H) = P(0)\frac{\Gamma_2^2}{\Gamma_2^2 + 4\Omega^2} - Q(0)\frac{2\Gamma_2\Omega}{\Gamma_2^2 + 4\Omega^2} \tag{2.9}$$

where $P(0)$ and $Q(0)$ are Stokes parameters of the studied spectral line in zero magnetic field.

The magnetic field dependence of the spectral line $P(H)$ is the superimposition of Lorentzian and dispersive contours. In order to illustrate the possibilities of the spectropolarimetric technique, let us examine the case of longitudial alignment with an axis oriented along the direction $\vec{n}$. In a zero magnetic field, the polarization moments which are of interest in (2.7) and (2.8), ${}^z\rho_{\pm 2}^{(2)}$, may be written as

$${}^z\rho_{\pm 2}^{(2)} = (\sqrt{6}/4)\,{}^{\vec{n}}\rho_0^{(2)} \sin^2\theta \exp(2i\varphi)$$

where θ and φ are angles determining the direction of the axis $\vec{n}$ in the detector frame (x, y, z). Expressions for Stokes parameters then take the form:

$$P(0) = P_0 \sin^2\theta \cos 2\varphi,\ Q(0) = P_0 \sin^2\theta \sin 2\varphi,$$

where P_0 is the degree of polarization for observations orthogonal to the alignment axis (1.30). In the case $\varphi = 0, \pi/2, \pi, ...$ when the vector $\vec{n}$ is in the plane XOZ or YOZ, the magnetic polarization signal has a Lorentzian shape which is characterized by the first Stokes parameter P. If the alignment axis is tilted at a certain angle to these planes, the non-zero Q Stokes parameter will appear as well as the dispersive part of the magnetic depolarization signal.

For $\varphi = \pi/4$, $3\pi/4$, ... the profile has a the purely dispersive shape. This means that tne alignment axis lies inclined at the angle $\varphi = \pi/4$ with respect to XOZ or YOZ planes. For arbitrary excitation, the pure Lorentzian shape shows that the alignment axis is in the XOZ or YOZ planes of the detector frame.

By these means, the profile of the registered magnetic field polarization dependence in the detector frame is determined by the orientation of the alignment axis. Therefore, analysing the signal profile in different parts of the discharge and for different orientations of magnetic field, it is possible to construct the spacial distribution of the alignment tensor and to determine the orientation of its axes. This is of great importance because it provides the possibility for determining the self-alignment mechanism. The second possibility is to measure the Stokes parameters in the linear spectrum. Moreover, the shape of the self-alignment signal enables us to obtain information about the relaxation constant Γ_2 and to measure the radiative lifetime and alignment destroying collisions cross-section /11/.

For an arbitrary oriented magnetic field, according to (2.6) and (2.7) we have:

$$P(H) = \left(w(JJ_1)/\rho_0^{(0)}\right)\left[\sum_{q'q''} D^{(2)}_{2q'} D^{(2)*}_{q''q'} F^{(2)}_{q''}/(\Gamma_2 + iq'\Omega) + \right.$$

$$\left. \sum_{q'q''} D^{(2)}_{-2q'} D^{(2)}_{q''q'} F^{(2)}_{q''}/(\Gamma_2 + iq'\Omega)\right] \tag{2.10}$$

where $D^{(2)}_{qq'}$ is the rotation matrix corresponding to the transition from the detector frame of reference to that with the OZ axis oriented along the magnetic field vector (Fig.13).

Multipole moments $F^{(2)}_{q''}$ are calculated in the detector frame of reference. Making some transformations of the expression (2.10) and using the definition of the rotation D-matrix:

$$D^{(\kappa)}_{-q-q'}(\alpha,\beta,\gamma) = (-1)^{q-q'} \exp\left[2iq\alpha + q'\gamma\right] D^{(\kappa)}_{qq'}; \; (-1)^q F^{(\kappa)}_{-q} = F^{(\kappa)*}_{q}$$
$$D^{(\kappa)}_{qq'}(\alpha,\beta,\gamma) = \exp -iq\alpha d^{(\kappa)}_{qq'}(\beta) \exp -iq\gamma$$

where $d^{(\kappa)}_{qq'}(\beta)$ are real functions of the β Euler angle, we have:

$$P(H)=\frac{2w(JJ_1)}{\rho_0^{(0)}}\sum_{q'q''}d^{(2)}_{2q'}d^{(2)}_{q''q'}\left[\frac{\Gamma_2 Re(F^{(2)}_{q''}e^{-i(2-q'')\alpha})}{\Gamma_2^2+q'^2\Omega^2}\right.$$

$$\left.+\frac{q'\Omega Im(F^{(2)}_{q''}e^{-i(2-q'')\alpha})}{\Gamma_2^2+q'^2\Omega^2}\right]$$

Say, for example, a magnetic field is oriented along the OX axis of the detector frame. The rotation Euler angles of the coordinate system are given by $\alpha=\gamma=0$, $\beta=\pi/2$. Using the D-matrix expression we have:

$$P(H)=\frac{w(JJ_1)}{\rho_0^{(0)}}\left[\left(\frac{1}{2}ReF_2^{(2)}+\frac{\sqrt{3}}{2\sqrt{2}}ReF_0^{(2)}\right)\frac{\Gamma_2}{\Gamma_2^2+4\Omega^2}-ImF_1^{(2)}\frac{2i\Omega}{\Gamma_2^2+4\Omega^2}\right.$$

$$\left.+2ReF_1^{(2)}\frac{\Gamma_2}{\Gamma_2^2+\Omega^2}+\left(\frac{3}{2}ReF_2^{(2)}-\frac{\sqrt{3}}{2\sqrt{2}}ReF_0^{(0)}\right)\frac{1}{\Gamma_2}+2ImF_2^{(2)}\frac{i\Omega}{\Gamma_2^2+\Omega^2}\right] \tag{2.11}$$

Hence, in this case the spectropolarimetric technique provides information about the $F_2^{(2)}$, $F_1^{(2)}$ and $F_0^{(0)}$ components of the excitation tensor. By changing the orientation of the magnetic field, one can determine all components of the excitation tensor for a given direction of observation. This can be easily realized in laboratory gas discharges.

For gas discharges,comparing this technique to direct spectropolarimetric methods, we must point out that the magneto-optical technique possesses a number of attractive features such as high sensitivity and high information capability. Another advantage of this method is connected to specific technical aspects. In order to measure the degree of polarization, it is not necessary to compensate for instrumental polarization. The useful signal in this case is the dependence of the spectropolarimetric characteristics on an external controllable parameter. Magnetic polarization technique enables also to perform some simple control experiments (Fig. 14) which help to establish the physical reason for self-alignment.

2.2.3 LABORATORY MAGNETIC POLARIZATION SPECTROMETER

The principle scheme of a magnetic polarization spectrometer for laboratory gas discharge studies is as follows:
The optical emission of the gas discharge is focused at the entrance plane of the spectroscopic device by a lens, with its optical axis parallel to the magnetic field. The radiation of the selected spectral line is divided by a

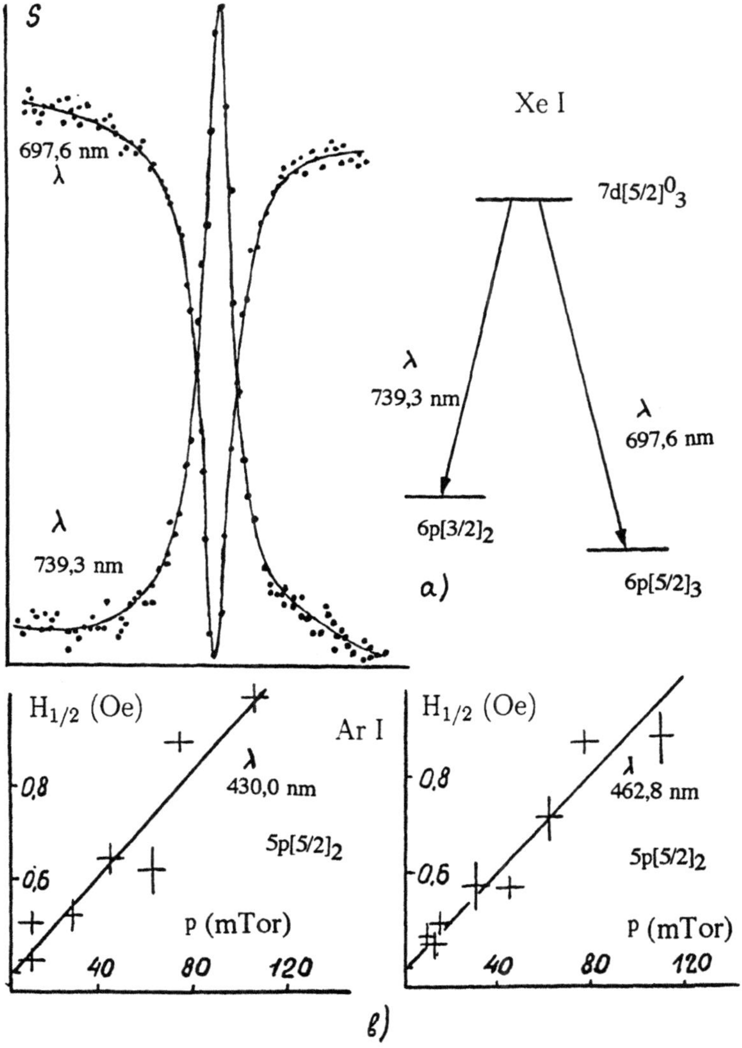

Figure 14. **Control experiments for measuring Stokes parameters of radiation emitted by a discharge plasma by the magneto-polarization technique: a) observation of different Hanle profiles for two transitions from one upper level J, b) comparison of pressure dependencies of widths of Hanle profiles**

polarization selector into two beams with orthogonal linear polarization direction parallel to the OX and OY axes of the detector frame. The intensities of these beams are registrered by two photomultipliers and the electric signals substracted.

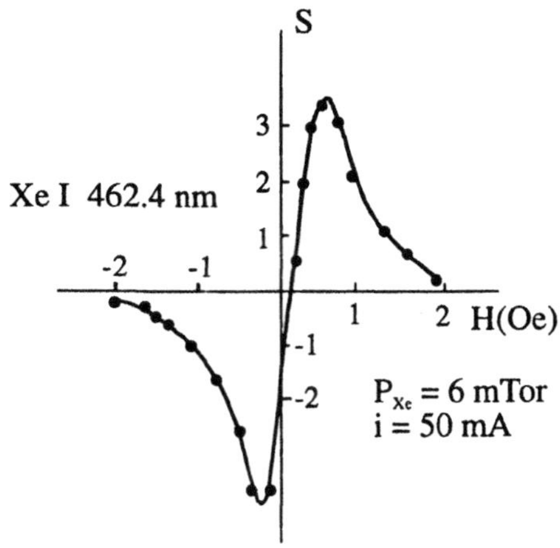

Figure 15. Self-alignment signal in 462.4 nm xenon line detected by magnetic field modulation technique

In order to detect weak magneto-optical dependencies, two methods of synchronic detection and signal storage are used /40/. In the first case, the magnetic field is modulated and the useful information is extracted from the modulated signal component synchronous with the modulated magnetic field. For this purpose, a differential amplifier connected to the synchronous detector is used. The reference sinusoidal signal for the synchronous detector is formed by the same generator which is used for the magnetic field modulation (Fig. 15).

This technique gives good results for the determination of the self-alignment mechanism and measurement of atomic constants. To determine the Stokes parameters and the measurement of the degree of polarization of emission of laboratory gas discharges, the scheme of a magnetic spectropolarimeter with storage of the signal, presented in Fig. 16, is always used. The gas discharge (2) is placed in the center of Helmholz coils (3) which produce an homogeneous magnetic field in the volume under study. Then the optical radiation is focused on the entrance slit of a monochromator (4). An interference polarizer (5) divides the emission of a given spectral line in two beams with orthogonal polarizations which are registered by two photomultipliers (6). The electric signals from the photomultipliers enter into the differential amplifier and are converted into digital form. The final signal (Fig. 17) is stored in a computor.

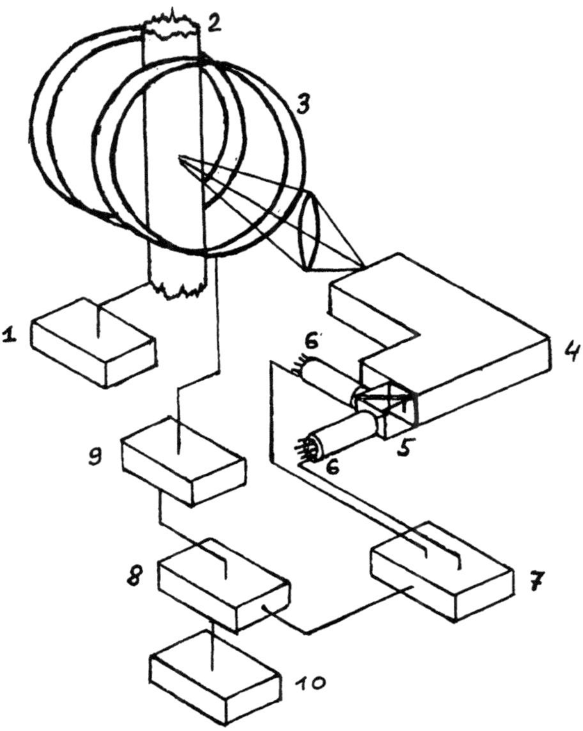

Figure 16. Scheme of magnetic spectropolarimeter with data storage

2.3 Derivation of local plasma parameters

A characteristic feature of every spectroscopic diagnostic method is that the measured parameters are averaged over the line of sight. The problem of deriving the local properties of every plasma layer from the integrated signal has been studied in details only with respect to the frequency distribution of lines emissivity /41, 42, 43, 44, 45/.

2.3.1 STOKES PARAMETERS AND POLARIZATION MOMENTS

In the traditional experimental arrangement for spectroscopic plasma diagnostics, the spatial distribution of the intensity in the plasma source is studied by projection in the plane of the entrance slit of the spectrograph. If the solid angles $\tilde{\tau}$ and $\tilde{\tau}'$ (Fig. 18) are small enough, the plasma homogeneous within these angles, and in the absence of absorption along the line of sight, the intensity received from a differentially small part dz of plasma in the frequency range $\omega, \omega + d\omega$ is given by:

$$dI_{\omega}d\omega = (4\pi)^{-1}\tilde{\varepsilon}_{\omega}(z)\tilde{\tau}S\,dz\,d\omega,$$

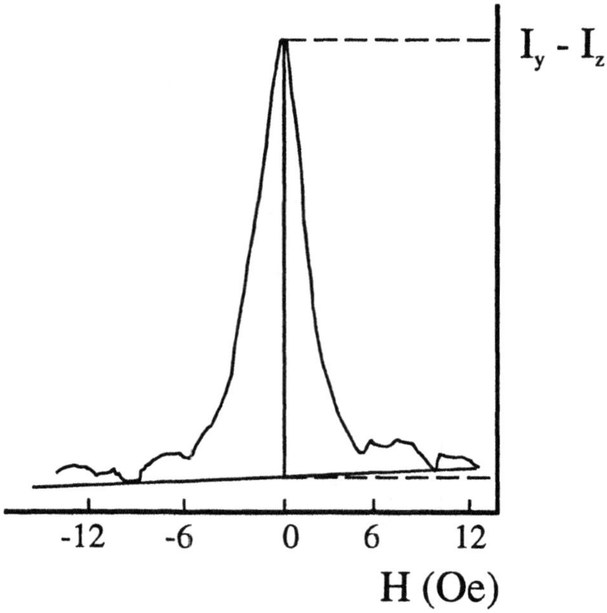

Figure 17. **Magnetic depolarization profile in the argon 763.5 nm line emited by a high frequency capacitive discharge at 100 MHz, 200 V, 8 mTorr**

where $\tilde{\varepsilon}_\omega(z)$ is the emissivity per unit volume and unit solid angle around z in the frequency range ω - $\omega + d\omega$, S is the area of the entrance slit or diaphragm /24/.

Due to absorption along the line of sight, the intensity received from the element dz outside the boundary of the plasma cell is:

$$dI_\omega d\omega = \frac{\tilde{\tau}}{4\pi}\tilde{\varepsilon}_\omega(z)S\ exp\left[-\int_{L_1}^{z} dz'\,\chi_\omega(z')\right]dzd\omega,$$

where $\chi_\omega(z)$ is the absorption coefficient at frequency ω, L_1 is the coordinate along the line of sight of the nearest plasma boundary (Fig. 18).

The total intensity will be expressed as:

$$I_\omega d\omega = \frac{S\tilde{\tau}}{4\pi}\int_{L_1}^{L_2} \tilde{\varepsilon}_\omega(z)\exp\left[-\int_{L_1}^{z} dz'\,\chi_\omega(z')\right]dzd\omega. \qquad (2.12)$$

In order to describe self-absorption in the plasma, a transparency parameter Π_ω is is used, which is defined as the ratio of ($I_{0\omega}$), the input intensity of the beam at the frequency ω, to I_ω the intensity of the output beam /44,46/:

$$I_\omega = \Pi_\omega\, I_{0\omega} = I_0\exp\left[-\int_{L_1}^{z} dz'\,\chi_\omega(z')\right]. \qquad (2.13)$$

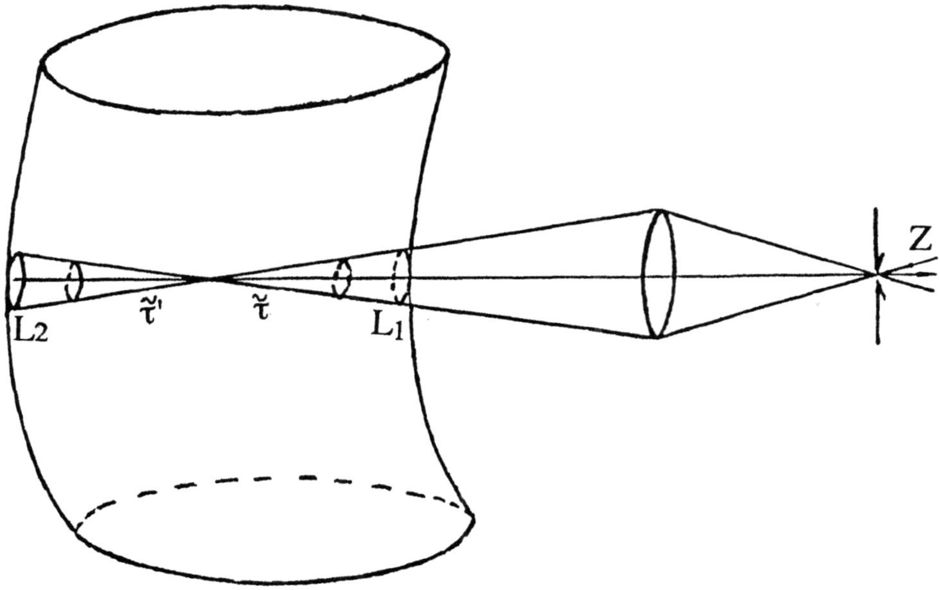

Figure 18. Scheme of experimental spectropolarimetric diagnostics of ionized gases

Assuming a low degree of excited states alignment, it is possible to generalize the expression (2.12) to obtain formulae for the density matrix of photons emitted within the solid angle $\tilde{\tau}$ and reaching the detector (Fig. 18):

$$ {}^{\omega}U_{\mu'\mu}d\omega = S\,d\omega \int_{L_1}^{L_2} {}^{\omega}U_{\mu'\mu}(z) \exp\left[-\int_{L_1}^{z} dz'\,\chi_\omega(z')\right] dz. $$

If the detector receives photons within the finite spectral range $\Delta\omega$ we have for the integral density matrix the following relation:

$$ U_{\mu'\mu} = \int_{\Delta\omega} d\omega\,{}^{\omega}U_{\mu'\mu} = S\int_{L_1}^{L_2} dz \int_{\Delta\omega} {}^{\omega}U_{\mu'\mu}(z) \exp\left[-\int_{L_1}^{z} dz'\,\chi_\omega(z')\right] d\omega. \tag{2.14} $$

Restricting ourselves to a Doppler broadened spectral absorption line, we may write:

$$ \chi_\omega = \chi_0 \exp\left(-\beta^2\right); \quad {}^{\omega}U_{\mu'\mu}d\omega = 2\pi^{-1/2}U_{\mu'\mu} \exp\left(-\beta^2\right) d\beta, \tag{2.15} $$

where

$$ \chi_0 = 2c(\pi M/2kT)^{1/2} $$

is the absorption coefficient at the center of a spectral line of frequency ω_0, M is the atomic mass, k the Boltzmann constant, c = light velocity, and χ

is the absorption coefficient

$$\beta = c\left[(\omega - \omega_0)/\omega_0\right](M/2kT)^{1/2}; \chi = 2\pi\omega N \mid \langle J \parallel d \parallel J_1 \rangle \mid^2 / (3c\hbar(2J_1+1)).$$

Thus the integral density matrix of optical radiation per unit volume corresponding to the spectral transition $J_1 \to J$ is expressed (1.24) by:

$$U_{\mu'\mu} = B(\omega)\tilde{\tau}\sum_{\kappa q}(-1)^{J+J_1+\mu}(2\kappa+1)^{1/2} \mid J \parallel \hat{d} \parallel J\rangle \mid^2$$

$$\begin{pmatrix} 1 & 1 & \kappa \\ \mu' & -\mu & q \end{pmatrix}\begin{Bmatrix} 1 & 1 & \kappa \\ J & J & J_1 \end{Bmatrix}\rho_q^{(\kappa)} \tag{2.16}$$

This expression is transformed to

$$U_{\mu'\mu} = \frac{2S}{\sqrt{\pi}}\int_{L_1}^{L_2} dz\, U_{\mu'\mu}(z)\int_{-\infty}^{\infty} d\beta\, \exp(-\beta^2)\exp\left[-\exp(-\beta^2)\int_{L_1}^{z} dz'\,\chi_0(z)\right]$$

$$= 2S\int_{L_1}^{L_2} dz\, U_{\mu'\mu}(z)S'(\xi), \tag{2.17}$$

where

$$\xi = \int_{L_1}^{z} dz'\,\chi_0(z'); \quad S' = \frac{1}{\sqrt{\pi}}\int_{-\infty}^{\infty} d\beta\, \exp(-\beta^2 - \xi e^{-\beta^2}). \tag{2.18}$$

The function S' is connected to the Ladenburg-Reche function $S(\xi)$ /24/ by the following relation:

$$S'(\xi) = d[\xi S(\xi)]/d\xi; \quad S(\xi) = \frac{1}{\xi\sqrt{\pi}}\int_{-\infty}^{\infty} d\beta\left[1 - \exp(-\xi e^{-\beta^2})\right]$$

For the simple case of a plasma homogeneous along the line of sight, $U_{\mu'\mu}$ and χ_0 are not z dependent and

$$U_{\mu'\mu} = 2SLU_{\mu'\mu}(z)S(\chi_0 L),$$

where $L = L_2 - L_1$ is the object dimension.

These relationships enable one to express the Stokes parameters I, P, Q in terms of the distribution of the polarization moments along the line of sight in the plasma:

$$I = \frac{2}{3}\sqrt{2J+1}B(\omega)\tilde{\tau}S \mid \langle J \parallel \hat{d} \parallel J_1\rangle \mid^2 I_0$$

$$P = I_0^{-1}\int_{L_1}^{L_2} dz\, P(z)\tilde{\varepsilon}_0(z)S'(\xi)$$

$$Q = I_0^{-1} \int_{L_1}^{L_2} dz\, Q(z)\tilde{\varepsilon}_0(z) S'(\xi) \tag{2.19}$$

where $I_0 = \int_{L_1}^{L_2} dz\, \tilde{\varepsilon}_0(z) S'(\xi)$

$$\begin{aligned} P(z) = &\ \tilde{\varepsilon}_0^{-1}(z) w(JJ_1) \left[\rho_2^{(2)}(z) + \rho_{-2}^{(2)}(z)\right] \\ Q(z) = &\ \tilde{\varepsilon}_0^{-1}(z) w(JJ_1)(-i) \left[\rho_2^{(2)}(z) - \rho_{-2}^{(2)}(z)\right] \\ \tilde{\varepsilon}_0(z) = &\ \rho_0^{(0)}(z) + \tfrac{2}{\sqrt{6}} w(JJ_1)\rho_0^{(2)}(z) \end{aligned}$$

are z-dependent Stokes parameters, and $\rho_q^{(\kappa)}(z)$ is the coordinate dependence of the polarization moments of the upper level of the studied spectral transition in the detector frame of reference.

2.3.2 AXIALLY SYMMETRIC IONIZED GAS ENTITIES

The problem of deriving the local plasma properties is of special practical significance for axially symmetric ionized gas entities. In the case of ordinary intensity spectroscopy, it is possible to construct a confined system of integral equations for the determination of the local optical parameters /44/. In order to examine the polarization characteristics of the linear radiation, let us direct the OY axis of the detector frame of reference (X, Y, Z) along the symmetry axis of the plasma entity (Fig. 19) . In the coordinate system $(X', Y', Z') : OZ' \parallel \vec{r}, (\widehat{OX, OX'}) = \varphi$, $\sin\varphi = x/r$ all the optical parameters of the axially symmetric plasma are dependent only on $r = |\vec{r}|$:

$$^{z'}\rho_q^{(\kappa)} = {}^{z'}\rho_q^{(\kappa)}(r); \quad \varepsilon = \varepsilon(r); \quad \chi_0 = \chi_0(r)$$

In order to obtain the expression of the polarization moments

$$^{z}\rho_{2+}^{(2)} = {}^{z}\rho_2^{(2)} + {}^{z}\rho_{-2}^{(2)}, \quad {}^{z}\rho_{2-}^{(2)} = {}^{z}\rho_2^{(2)} - {}^{z}\rho_{-2}^{(2)}$$

in the detector frame of reference, we use the rotational transformation:

$$\begin{aligned} ^{z}\rho_{2+}^{(2)} = &\ \textstyle\sum_{q=-2}^{2} \left[D_{q2}^{(2)}(z'z) + D_{q-2}^{(2)}(z'z)\right]^* \, {}^{z'}\rho_q^{(2)} \\ ^{z}\rho_{2-}^{(2)} = &\ \textstyle\sum_{q=-2}^{2} \left[D_{q2}^{(2)}(z'z) - D_{q-2}^{(2)}(z'z)\right]^* \, {}^{z'}\rho_q^{(2)} \end{aligned}$$

where $D_{q'q}^{(2)}(z'z)$ are Wigner matrix elements corresponding to the rotation of the coordinate system by an angle φ around the OY axis. As a result we have:

$$\begin{aligned} ^{z}\rho_{2+}^{(2)} = &\ (1 + \cos\varphi^2)/2\rho_{2+}^{(2)} + \sqrt{\tfrac{3}{2}} \sin^2\varphi \, {}^{z'}\rho_0^{(2)} + \tfrac{1}{2}\sin 2\varphi \, {}^{z'}\rho_{1-}^{(2)} \\ ^{z}\rho_{2-}^{(2)} = &\ \cos\varphi \, {}^{z'}\rho_{2-}^{(2)} + \sin\varphi \, {}^{z'}\rho_{1+}^{(2)} \end{aligned} \tag{2.20}$$

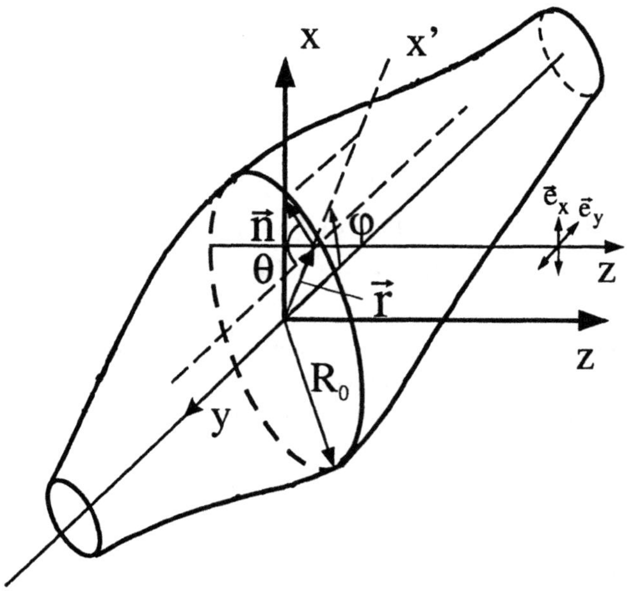

Figure 19. Representation of the axis of coordinates used for the calculation of the polarization of a axially symmetric plasma emission

where

$$ {}^{z'}\rho^{(2)}_{1+} = {}^{z'}\rho^{(2)}_{1} + {}^{z'}\rho^{(2)}_{-1}, \; {}^{z'}\rho^{(2)}_{1-} = {}^{z'}\rho^{(2)}_{1} - {}^{z'}\rho^{(2)}_{-1} $$

Polarization moments ${}^{z'}\rho^{(2)}_{1-}$, ${}^{z'}\rho^{(2)}_{2-}$ are asymmetric with respect to reflection in the YZ' plane, therefore for an axially symmetric entity, ${}^{z'}\rho^{(2)}_{1-} = {}^{z'}\rho^{(2)}_{2-} = 0$. Making the substitution of variables $r = \sqrt{x^2 + z^2}$, using (2.20), and assuming a low degree of alignment of atoms in the object ($\rho^{(2)}_q \ll \rho^{(0)}_0$), we obtain the following expressions for the distributions of Stokes parameters across the image of the studied object, :

$$
\begin{aligned}
P(x) &= \frac{w(JJ_1)}{I_0(x)} \int_x^R \frac{dr\, r}{\sqrt{r^2-x^2}} \left[\frac{2-x^2/r^2}{2} Z'\rho^{(2)}_{2+}(r) + \sqrt{\frac{3}{2}\frac{x^2}{r^2}}\, {}^{z'}\rho^{(2)}_0(r)\right] \\
&\quad [S'(\xi_{rR}) + S'(\xi_{xr} + \xi_{xR})] \\
Q(x) &= \frac{w(JJ_1)}{iI_0(x)} \int_x^R \frac{dr\, r}{\sqrt{r^2-x^2}}\, {}^{z'}\rho^{(2)}_0(r)\, [S'(\xi_{rR}) + S'(\xi_{xr} + \xi_{xR})] \qquad (2.21) \\
I_0(x) &= \int_x^R \frac{dr\, r}{\sqrt{r^2-x^2}}\, \rho^{(0)}_0(r)\, [S'(\xi_{rR}) + S'(\xi_{xr} + \xi_{xR})]
\end{aligned}
$$

where $\xi_{r_1 r_2} = \int_{r_1}^{r_2} \frac{dr'\, r'}{\sqrt{r'^2-x^2}} \chi_0(r')$ and R is the radius of the studied cross section of the plasma.

These equations may be regarded as a system of integral equations with respect to the unknown functions $\chi_0(r), \rho_0^{(0)}(r), \; {}^{z'}\rho_q^{(2)}(r)$. Using the expression (2.18) for $S'(\xi)$, the expression for $I_0(x)$ (2.21) takes the form:

$$\begin{aligned} I_0(x) &= \tfrac{2}{\sqrt{\pi}} \int_{-\infty}^{\infty} d\beta \, e^{-\beta^2} exp\left[-e^{-\beta^2} \int_x^R \tfrac{dr\, r\chi_0(r)}{\sqrt{r^2-x^2}}\right] \int_x^R \tfrac{dr\, r\rho_0^{(0)}(r)}{\sqrt{r^2-x^2}} \\ & ch\left[e^{-\beta^2} \int_x^r \tfrac{dr'\, \chi_0(r')r'}{\sqrt{r'^2-x^2}}\right] \end{aligned} \qquad (2.22)$$

For an axially symmetric plasma, if the transparency in the center of the spectral line, i.e.

$$\Pi_{\omega_0}(x) = exp\left[-2\int_x^R \frac{dr\, r\chi_0(r)}{\sqrt{r^2-x^2}}\right] = \exp(-2\xi_{xR}) \qquad (2.23)$$

is more than 0.5, the expressions for the Stokes parameters may be simplified as /44/:

$$P(x) = \frac{w(JJ_1)}{I'(x)} \int_x^R \frac{dr\, r}{\sqrt{r^2-x^2}} \left[\frac{2 - x^2/r^2}{2} \; {}^{z'}\rho_{2+}^{(2)} + \sqrt{\frac{3}{2}\frac{x^2}{r^2}} \; {}^{z'}\rho_0^{(2)}(r)\right]$$

$$Q(x) = \frac{w(JJ_1)}{iI'(x)} \int_x^R \frac{dr\, r}{\sqrt{r^2-x^2}} \; {}^{z'}\rho_{1+}^{(2)}(r)$$

$$I_0(x) = 2S'(\xi_{xR})I'(x); \;\; I'(x) = \int_x^R \frac{dr\, \rho_0^{(0)}(r) r}{\sqrt{r^2-x^2}} \qquad (2.24)$$

For longitudinal alignment, these equations may be transformed using the P_0 parameter, which is the degree of polarization of the optical radiation observed at a right angle with respect to the alignment axis. Using the representation by the polarization moments (1.29) and variables given in Fig. 19, we may write expressions for the local values of the Stokes parameters:

$$\begin{aligned} P(z) &= \; P_0(z)(\sin^2\theta \sin^2\varphi - \cos^2\theta) \\ Q(z) &= \; P_0(z)\sin 2\theta \sin\varphi, \end{aligned}$$

where θ and φ are angles defining the direction of the alignment axis in the laboratory frame of reference. Using the relationships $\sin\varphi = x/r$ we have for the Stokes parameters:

$$P(x) = [I'(x)]^{-1} \int_x^R dr \frac{r\rho_0^{(0)} P_0(r)}{\sqrt{r^2-x^2}} \left[\left(\frac{x^2}{r^2}\right)\sin^2\theta - \cos^2\theta\right]$$

$$Q(x) = [I'(x)]^{-1} \int_x^R dr \frac{r\rho_0^{(0)} P_0(r)}{\sqrt{r^2-x^2}} \sin 2\theta \qquad (2.25)$$

These expressions may be applied to a number of practical cases. For a spherical object characterized by the radial symmetry of the alignment distribution, we have: $Q = 0$, $\cos\theta = 0$, $\sin\theta = 1$ and the integral representation of the degree of polarization takes the form:

$$P(x) = \left[I'(x)\right]^{-1} \int_x^R \frac{\rho_0^{(0)}(r) P_0(r) x^2}{r\sqrt{r^2 - x^2}} dr \tag{2.26}$$

For an axially symmetric object, we have $\theta = 0$ and $Q = 0$. In this case the integral polarization optical characteristics are described by equations of Abel type:

$$I_0 = 2S'(\xi_{xR}) \int_x^R \frac{dr\, r\rho_0^{(0)}(r)}{\sqrt{r^2 - x^2}}$$

$$P(x) = \left[I'(x)\right]^{-1} \int_x^R \frac{dr\, r\rho_0^{(0)}(r) P_0(r)}{\sqrt{r^2 - x^2}}. \tag{2.27}$$

For the case of an uniform distribution of the degree of polarization across the plasma image, $P(x) = P = const$, the volume distribution of alignment is constant, $P_0(r) = P$. For an uniform volume distribution of the population of the upper state of the transition $\rho_0^{(0)}(r) = \rho_0^{(0)} = const$, according to (2.27), the distribution of intensity and the degree of polarization take the form:

$$I_0(x) = 2S'(\xi_{rR})\rho_0^{(0)}\sqrt{R^2 - x^2}; \;\; P(x)\sqrt{R^2 - x^2} = \int_x^R \frac{dr\, P_0(r) r}{\sqrt{r^2 - x^2}}$$

If the $P(x)$ distribution may be approximated by a polynomial of even powers, the $P_0(r)$ dependence will be represented by a polynomial of the same powers. For many cases it is possible to use a polynomial of the second and fourth power:

$$P(x) = A_0 + A_1(x/R)^2 + A_2(x/R)^4$$

To this function corresponds the polynomial $P_0(r) = B_0 + B_1(x/R)^2 + B_2(x/R)^4$.... with coefficients $B_0 = A_0 - A_1/2 - A_2/8$, $B_1 = 3A_1/2 - 3A_2/4$, $B_2 = 15A_2/8$. This approach by approximating polynomials was used for polarization measurements in a spherical high frequency discharge on 626.6 nm *NeI* line at 60 mTorr (Fig. 20). Additional measurements showed that $\Pi_{\omega_0} = 1$ and $I_0(x) = \alpha\sqrt{R^2 - x^2}$ (α =const). So the distribution of the excited atoms in the discharge may be considered to be uniform. A least square procedure was used to estimate the coordinate dependence of the observed polarization

$$P(x) = 3.0 \cdot 10^{-3} \left[1 - 0.85(x/R)^2 + 0.21(x/R)^4\right]$$

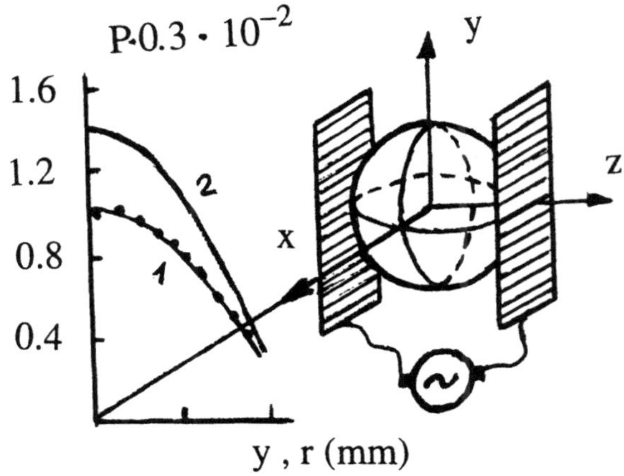

Figure 20. Derivation of the spatial dependence of the polarization from spectropolarimetric measurements in 626.6 nm neon line emitted by a spherical high frequency capacitive discharge at 100 MHz and 60 mTorr

(Fig. 20). Using the relationship between the coefficients of the two polynomials $P(x)$ and $P_0(r)$, we have

$$P_0(r) = 3.0 \cdot 10^{-3} \left[1.4 - 1.43(r/R)^2 + 0.39(r/R)^4\right].$$

Comparing the spatial dependencies of $P_0(r)$ and $P(x)$, we may conclude that polarization in the center of the discharge, $P_0(0)$, characterizes the degree of alignment in the center of discharge and that it is given by the formula $P_0(0) = 1.4 \cdot P(0)$ /42, 43/. This result shows that for diagnostic applications, it is important to account for volume effects and to take into account the real dimensions of the ionized gas entity.

CHAPTER 3

POLARIMETRIC SENSING OF THE POSITIVE COLUMN IN A D.C DISCHARGE

The positive column plasma of a d.c. discharge is a laboratory low temperature plasma object, which has been studied extensively by intensity optical spectroscopic methods and by the electric probe technique. The properties of a d.c. discharge are well known, which make this object valuable for checking new diagnostic techniques. Therefore, the diagnostics capabilities of polarization spectroscopy can then be tested by comparison of the data obtained using independant methods of probing d.c. discharges.

3.1 Experimental characteristics of noble gas self-alignment

Studies of self-alignment in the positive column plasma of a gas discharge have been carried out using the magnetic spectropolarimetric technique. Applied to a laboratory discharge plasma, this method is very useful for measuring Stokes parameters in different spectral lines, to study the distribution of self-alignment on different energy levels, and to examine the details of the magnetic spectropolarimetric profiles for different experimental conditions.

3.1.1 CASCADING SELF-ALIGNMENT

Helium /47, 48/. Self-alignment of singlet and triplet helium states have been observed using a number of spectral lines in cylindrical positive column discharges for 0.1 – 0.7 Torr and 20 – 70 mA (Fig. 21). Study of the magnetic depolarization effects showed that self-alignment of helium states is induced by a direct elementary excitation process in the plasma that is described in chapter 1. The linear pressure broadening of the magnetic depolarization signals for $3, 4, 5\,^1D_2$ levels indicated that atom - atom collisions are responsible for the depolarization of excited states. Extrapolation to zero pressure (Fig. 22) and estimates of relaxation constants for $3, 4\,^1P_1$ levels showed that the resonance radiation self-absorption on $3, 4\,^1P_1 - {}^1S_0$ transitions plays a very important role in the process of self-alignment relaxation in the plasma /22/.

Argon /49, 50/. The self-alignment effect in the heavy noble gases appears to be more complicated. Self-alignment of $2p$ levels (see generalized level scheme Fig. 23) was observed at pressure 0.02 – 2 Torr and for discharge currents in the range 20 – 80 mA. Pressure dependence of the width of selfalignment signals (Fig. 24) and estimates for alignment relaxation con-

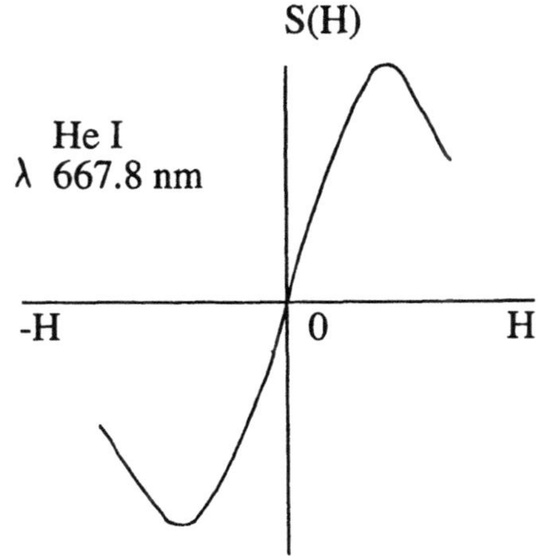

Figure 21. Self-alignment signal shape in a d.c. discharge detected by magnetic field modulation technique in the 667.8 nm helium line

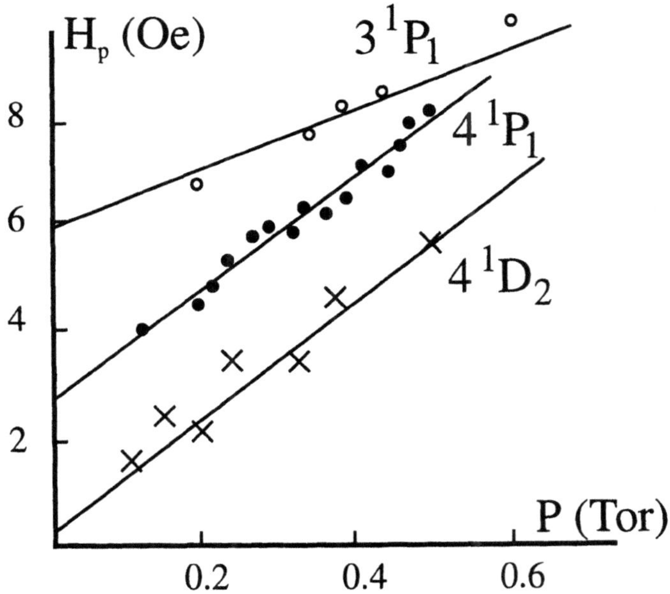

Figure 22. Extrapolation to zero pressure in the positive column of d.c. discharge of self-alignment signal widths for $3, 4^1P_1$ and 4^1D_2 helium levels

stants indicated that, for a pressure exceeding 0.3 Torr, the self-alignment is induced by radiation self-absorption in $1s - 2p$ transitions, and that the relaxation is due to atom - atom collisions. Some peculiarities have been detected for pressures less than 0.3 Torr. Pronounced nonlinearities of the pressure dependencies of the widths of the self-alignment signals have been

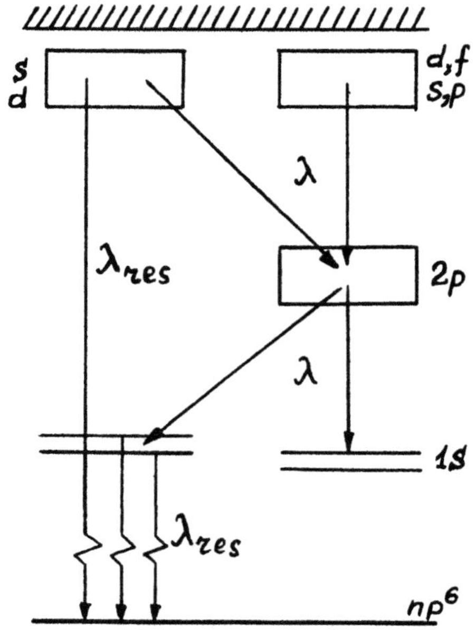

Figure 23. **Generalized scheme of levels and spectral transitions in noble gases**

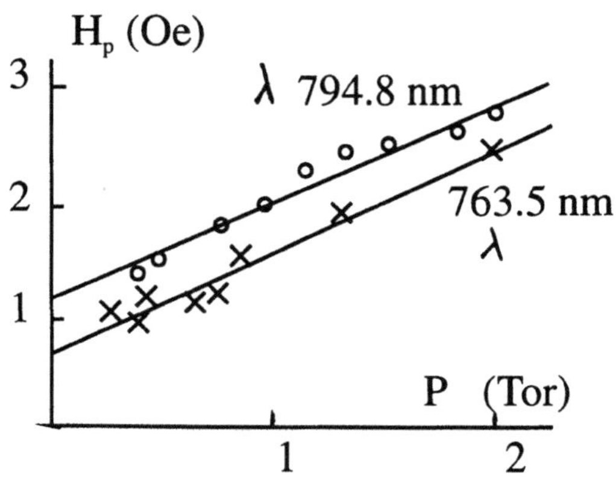

Figure 24. **Pressure variation in d.c. discharge, for pressures exceeding 0.3 Torr, of self-alignment signal widths of 2p argon levels**

observed for all $1s - 2p$ spectral transitions (Fig. 25). These nonlinearities could not be explained as a result of radiation self-absorption in $1s - 2p$ transitions /18, 22/. This conclusion has been confirmed by estimates of relaxation constants and by supplementary experiments /22, 50/. Besides, complicated specific two-component magnetic depolarization signal shapes for $2p$ argon levels have been observed in low pressure range (0.02 – 0.2 Torr). The broad component was determined by the characteristic relax-

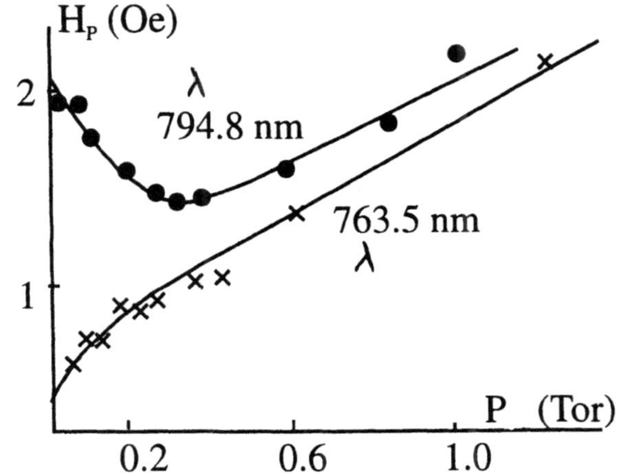

Figure 25. Pressure variation in d.c. discharge, for low pressures, of self-alignment signal width of 2p argon levels

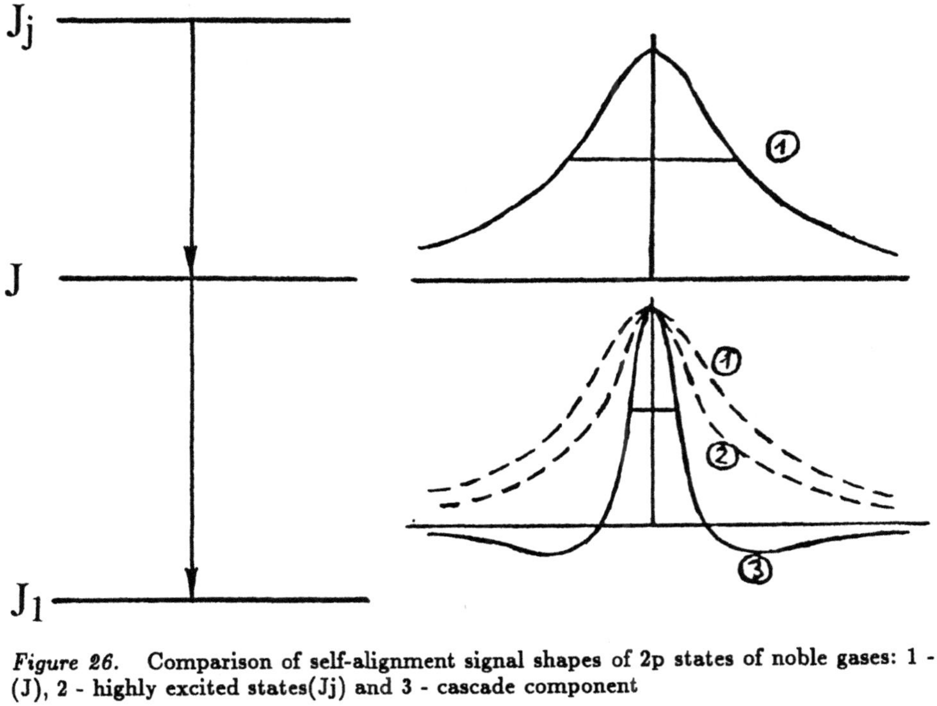

Figure 26. Comparison of self-alignment signal shapes of 2p states of noble gases: 1 - (J), 2 - highly excited states(Jj) and 3 - cascade component

ation time of the upper $2p$ level. The relative contribution of the narrow component increased when the pressure approached zero and for the lowest possible pressures only the narrow component was observed. This narrow component was attributed to self-alignment cascading processes to the levels belonging to the $3p^5 4p$ configuration.

Cascade transitions due to the isotropy of the spontaneous emission transfer the polarization moments of the higher excited levels to the studied ones, with no change of the rank κ. This means that the alignment of the upper levels may be transformed only into the alignment of the lower ones (Fig. 26). To describe the cascade process one should add the operator $\hat{F}'$ to the right-hand side of the equation for the statistical operator (1.26). Within the $J - m$, representation, the matrix elements of $\hat{F}'$ (1.31) are given by:

$$F'_{m'm} = \sum_{i}\sum_{\mu} B(\omega_j) \int_{4\pi} d\vec{n} \langle Jm' \mid (\hat{\vec{d}} - (\vec{n}\hat{\vec{d}})\vec{n})_{\mu} \hat{\rho}_{Jj} (\hat{\vec{d}} - (\vec{n}\hat{\vec{d}})\vec{n})^{+}_{\mu} \mid Jm\rangle \tag{3.1}$$

where ω_j is the frequency of $J_j - J$ transition , $\mid J_j m_j\rangle$ are wave functions and $\hat{\rho}_{Jj}$ is the statistic operator of these states. This expression is transformed to:

$$F'_{m'm} = 4\pi \sum_{j} B(\omega_j) \sum_{\mu\nu\nu'} P_{\mu\nu\mu'\nu'} \langle Jm \mid d_{\nu} \hat{\rho}_{Jj} d^{+}_{\nu'} \mid Jm'\rangle \tag{3.2}$$

where $P_{\mu\nu\mu'\nu'}$ is the operator reflecting the transversity of the electromagnetic wave.

Using polarization moment formalism and the Wigner - Eckart theorem we have:

$$F'^{(\kappa)}_{q} = \sum_{j} (-1)^{J+J_j+\kappa+1} \frac{8\pi}{3} B(\omega_j) \mid J \parallel \hat{d} \parallel J_j\rangle \mid^2 \left\{ \begin{array}{ccc} J_j & J_j & \kappa \\ J & J & 1 \end{array} \right\} \rho^{(\kappa)}_{q}(J_j)$$

Within the reference frame with the quantization axis oriented along the magnetic field direction, the equation for the polarization moment evolution (2.5) takes the form:

$$\dot{\rho}^{(\kappa)}_{q}(J) = (iq\Omega - \Gamma_{\kappa})\rho^{(\kappa)}_{q}(J) + F^{(\kappa)}_{q}(J)$$
$$+ \tfrac{8\pi}{3} \textstyle\sum_j (-1)^{J+J_j+\kappa+1} B(\omega_j) \mid J \parallel \hat{d} \parallel J_j\rangle \mid^2 \left\{ \begin{array}{ccc} J_j & J_j & \kappa \\ J & J & 1 \end{array} \right\} \rho^{(\kappa)}_{q}(J_j) \tag{3.3}$$

where Ω is the Larmour frequency for the state J.

For the polarization moments of the upper levels, $\rho_q^{(\kappa)}(J_j)$, we may write in turn the following evolution equation:

$$\dot{\rho}_q^{(\kappa)}(J_j) = (iq\Omega_j - \Gamma_{j\kappa})\rho_q^{(\kappa)}(J_j) + F_q^{(\kappa)}(J_j), \tag{3.4}$$

where $\Omega_j, \Gamma_{j\kappa}, F_q^{(\kappa)}(J_j)$ are the Larmor frequency, the relaxation constant and the multipole moment of the excitation tensor for the state J.

For stationary conditions,

$$\dot{\rho}_q^{(\kappa)}(J) = 0, \dot{\rho}_q^{(\kappa)}(J_j) = O,$$

the solution of the equation (3.3) takes the form:

$$\begin{aligned} \dot{\rho}_q^{(\kappa)}(J) &= \frac{F_q^{(\kappa)}(J)}{\Gamma_\kappa + iq\Omega_J} \\ &+ \frac{8\pi}{3}\sum_j(-1)^{J+J_j+\kappa+1}B(\omega_j)\left\{\begin{array}{ccc} J_j & J_j & \kappa \\ J & J & 1 \end{array}\right\} \\ &[F_q^{(\kappa)}(J_j) \mid \langle J \parallel d \parallel J_j\rangle \mid^2]/[(\Gamma_{j\kappa} - iq\Omega_j)(\Gamma_\kappa - iq\Omega)] \end{aligned} \tag{3.5}$$

Using these relations it is easy to calculate the dependence of the linear polarization of emission on the magnetic field strength for a three-levels atome scheme. Restricting ourselves to the simpliest case when the coherence is not induced on the studied level J $(F_q^{(\kappa)}(J) = F_0^{(0}\delta_{\kappa 0})$ and the alignment of this state appears only in the cascade transitions for which we calculated the magnetic dependence, the profile for 2p states of noble gases, $(\tau \ll \tau_j, g, g_j > 0)$, is much narrower than the signal profile corresponding to the direct coherent excitation of the level under study (Fig. 26). If self-alignment is induced to all levels, the signal will be a sum of the contributions reflecting the transfer of self-alignment from all levels involved. Such two-component signals were observed experimentally for $2p$ levels of noble gases.

Cascading of alignment from highly excited states is the reason for the nonlinear pressure dependence of the width of the signal on 2p argon levels which was observed in the positive column plasma of gas discharge at low pressures (Fig. 25).

Krypton /51/. The self-alignment effect was most pronounced on $2p$ levels of krypton and it was possible to carry out detailed study of the signal shape in a wide range of pressures (0.005 – 0.3 Torr). A simple Lorentzian shape was observed for the pressure range 0.05 – 0.3 Torr. For the majority of $2p$ levels at pressures 0.005 – 0.05 Torr, a complicated two-component signal profile was detected (Fig. 27, 28). The relative contribution of the narrow one increased when the pressure was diminished. The signal was represented

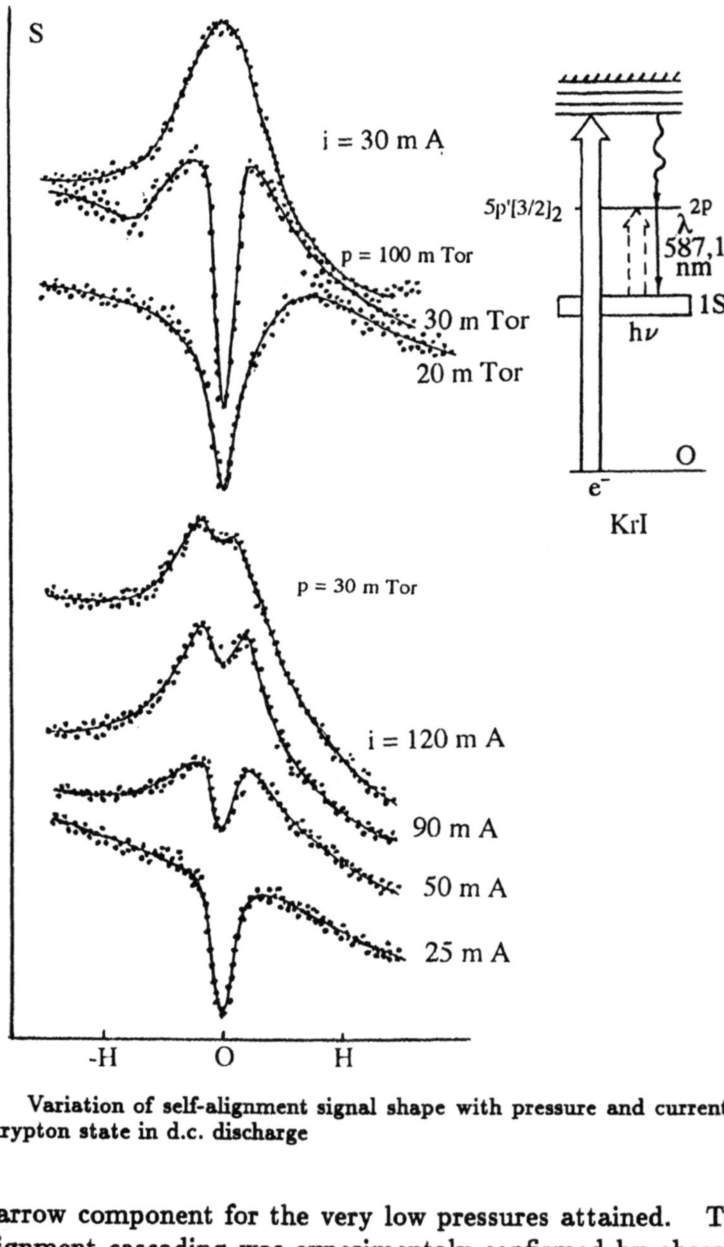

Figure 27. **Variation of self-alignment signal shape with pressure and current for the 5p'[3/2]$_2$ krypton state in d.c. discharge**

by the narrow component for the very low pressures attained. The role of self-alignment cascading was experimentaly confirmed by observations on 5p $[5/2]_3$ krypton level (Fig. 29). The narrow component was detected for the same discharge conditions as the strong self-alignment signals on the transitions from the highly excited 5d $[7/2]_3$, 6d $[7/2]_4$, 7d $[7/2]_4$ states on the lines 792.8, 645.6, 583.2 nm terminating on the level $5p[5/2]_3$.

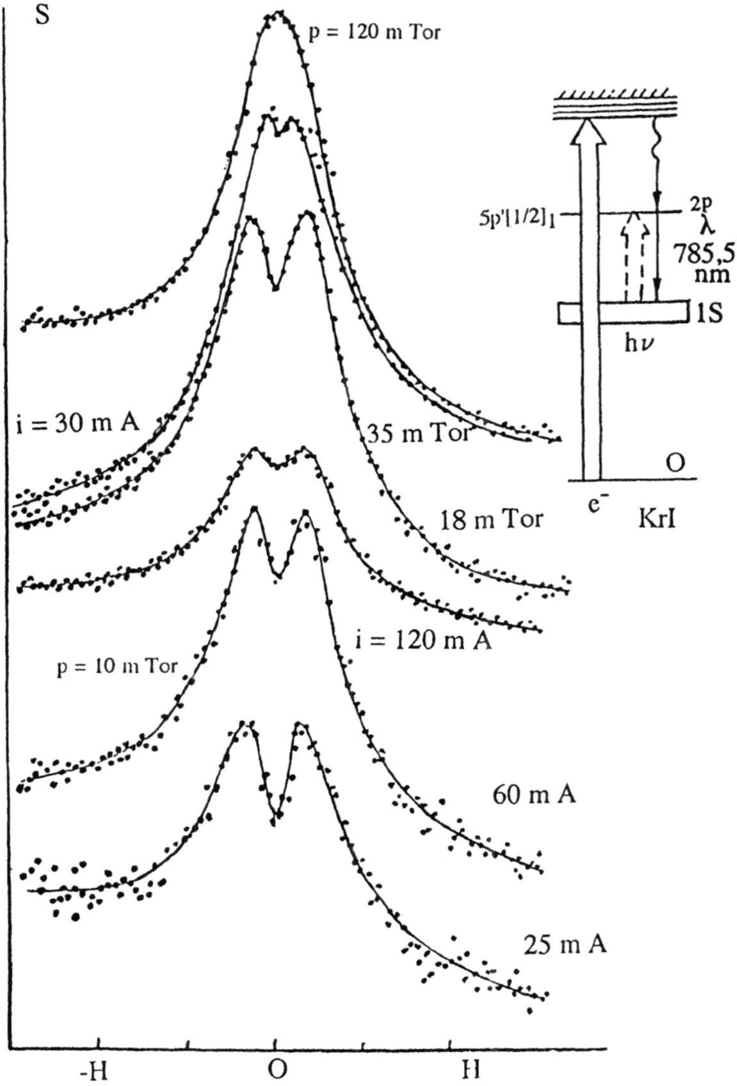

Figure 28. **Variation of the self-alignment signal shape with pressure and current for the 5p'[1/2]$_1$ krypton state in d.c. discharge**

For pressures greater than 0.05 Torr, the self-alignment signal of $5p[5/2]_3$ level became Lorentzian due to effective collisional destruction of the self-alignment of the highly excited levels. The discharge current dependencies (Fig. 27, 28, 30) also corroborated this conclusion. The observed increase of the relative contribution of the broad signal was in accordance with the current growth of the populations of the intermediate metastable $1s$ states

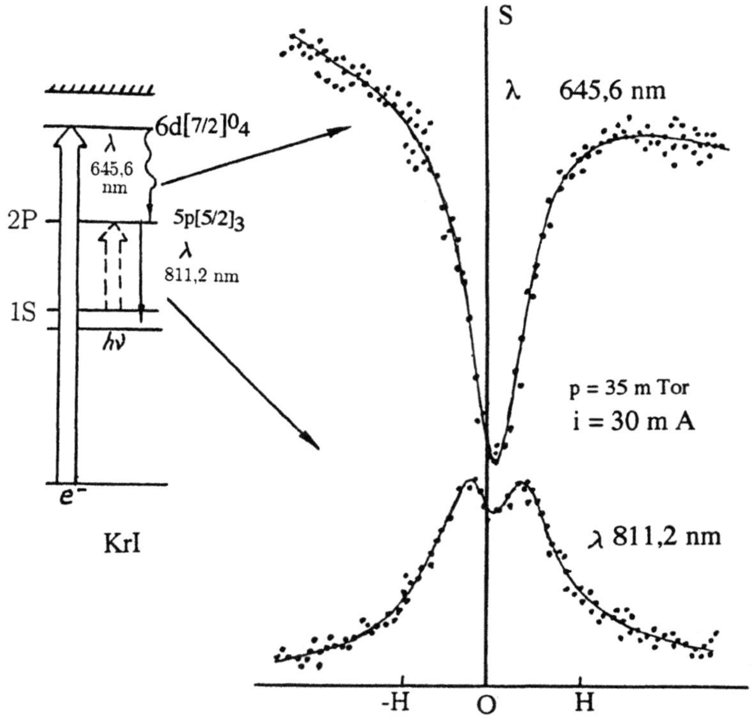

Figure 29. **Self-alignment signals detected on 2p level and on optically connected highly excited states of krypton in d.c. discharge**

and the corresponding increase of the $2p$ states self-alignment efficiency, due to the $1s-2p$ radiation self-absorption. At the same time, the electron impact-destroying cross-section for the highly excited levels is much higher than for $2p$ levels.

Xenon /52/. The self-alignment effect on $2p$ and $3p$ xenon levels was observed at pressures lower than 100 mTorr and for a discharge current in the range $10-100$ mA. Cascading self-alignment from highly excited states has been also detected in the range of the low pressures as for lighter noble gases.

As a result, observations of the self-alignment of $2p$ states of the noble gases indicated that a strong self-alignment of the highly excited atomic states at low pressures was induced and that this effect had a specific physical reason.

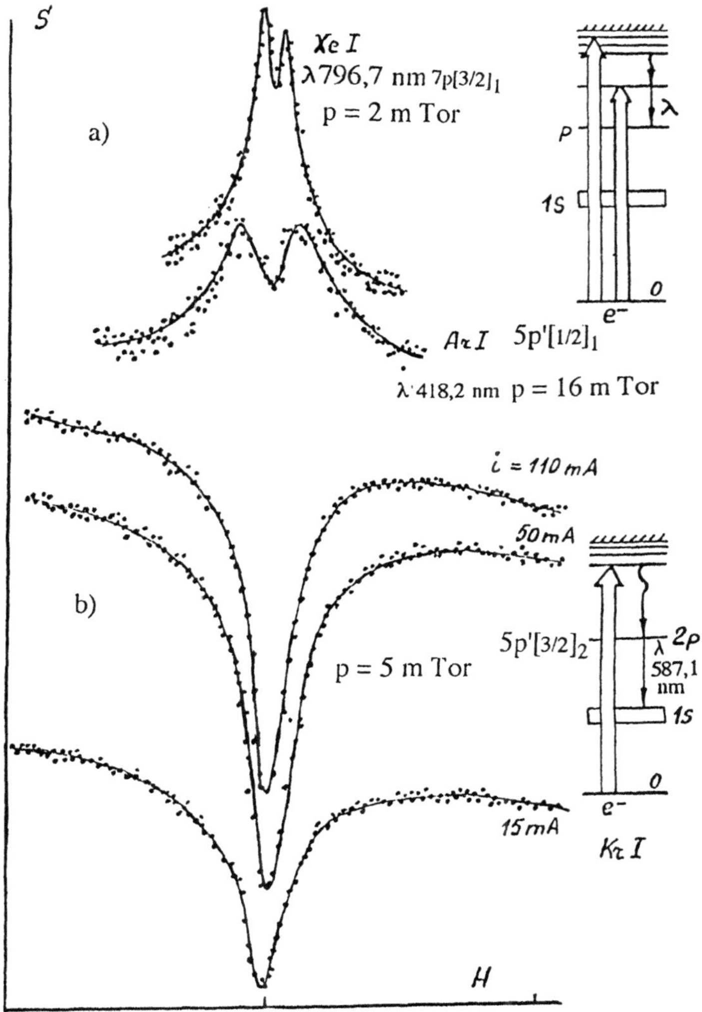

Figure 30. **Cascade effects in a positive column plasma at very low pressures of noble gases: a) cascade self-alignment signals for highly excited argon and xenon states, b) current dependent self-alignment signals for krypton states**

3.1.2 SELF-ALIGNEMENT OF HIGHLY EXCITED STATES

The self-alignment of the highly excited states of noble gases has been examined directly on the spectral transitions originating from this group of levels.

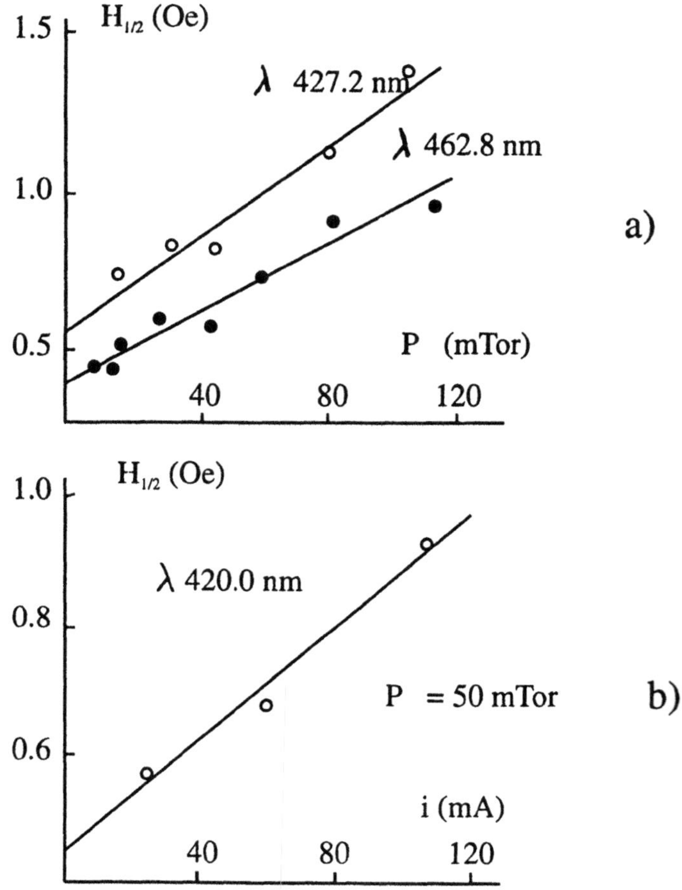

Figure 31. **Pressure (a) and current (b) dependencies of self-alignment signal widths of $3p^5 5p$ argon levels in the positive column plasma of a d.c. discharge**

Neon /53/. In neon a pronounced self-alignment effect of *nd* and *n's* states was observed in the transitions terminating on $2p$ levels within the pressure range 0.1 – 0.5 Torr and for a discharge current of 10 – 70 mA.

Argon /54/. Strong selfalignment was detected in argon on levels not only optically connected to $2p$ levels, for example $5p$ levels. Self-alignment of the block of the highly excited states was also detected for pressures lower than 0.1 Torr. The profile broadening by pressure and by a discharge current was studied in detail for these levels (Fig. 31).

Krypton /55, 56/. For the levels belonging to $4p^5 6p$ configuration, self-alignment effect was observed for pressures 0.05 – 0.14 Torr and a discharge current of 20 – 120 mA. For $nd, n's$, and $n''f$ configurations this effect also detected for the narrow range of very low pressures 0.005 – 0.05 Torr. The signal shape was Lorentzian with a width determined by the alignment relaxation constant of the upper level.

Xenon /55, 57/. Experimental observations in a xenon discharge confirmed that self-alignment of the highly excited levels in low pressure d.c. discharges is a general feature. The most pronounced effect was detected on $7p$, $7d$ and $8d$ electron configurations in the pressure range 2 – 70 mTorr and for a discharge current 10 – 70 mA. The self-alignment of $7d$, $4f$, $8s$, $8p$ and $8d$ levels was observed within the narrow pressure range 15 – 20 mTorr for a discharge current of 60 mA.

Cascading between the highly excited states of noble gases was also detected for the $3p$ levels, for example for the $5p'[1/2]_1$ level of argon (Fig. 30) and for some levels of xenon, as a two-components signal. The signal shape in this case was not pressure dependent. It means that the self-alignment of all highly excited states is induced by the same process which is most effective for low gas pressures.

The selective self-alignment of highly excited resonance states may be explained within a model of anisotropic self-absorption of radiation at the plasma boundary. Simple estimates of the photon mean free-path at the central frequency of the Doppler broadened line confirm this mechanism. Nevertheless, self-absorption of radiation cannot explain the self-alignment of the block of nonresonant states. The study of magnetic profiles of the polarization signals excluded a collisional transfer of self-alignment from the resonance states. On the other hand, comparison of the degree of self-alignment of nonresonance levels with resonance ones for the heavy noble gases showed that electron impact inelastic collisions /10, 55/ must be taken into consideration for explanation of the highly excited state self-alignment.

3.2 Electron impact self-alignment

The best objects for the study of electron impact self-alignment at low pressures are the highly excited states of noble gases in the positive column plasma of d.c discharge. First, we shall analyse experiments which aimed to demonstrate the collisional origin of self-alignment and examine its specific features.

3.2.1. EVIDENCE FOR ELECTRON IMPACT SELF-ALIGNMENT

The first experiment to determine the origin of self-alignment in a positive column plasma was performed in a specially shaped discharge vessel /58/. It is known that the distribution of self-alignment arising from the anisotropy of the self-absorption is connected to the shape of the plasma entity. The intensity of resonance radiation propagating in a plasma along a direction determined by the vector $\vec{n}$ is expressed as:

$$I(\vec{n}) = \tilde{\alpha} \int_0^{L(\vec{n})} dl\, \tilde{\varepsilon}(l) \exp\left[-\int dl'\, \chi_0(l')\right], \tag{3.6}$$

where $\tilde{\varepsilon}(l)$ is the distribution of the local radiation emissivity of a plasma as a function of the distance l in the direction defined by the vector $\vec{n}$ and varying from 0 to $L(\vec{n})$ in the plasma. $\chi_0(l)$ is the absorption coefficient distribution and $\tilde{\alpha}$ is constant.

If the plasma is uniform: $\tilde{\varepsilon}(l) = \tilde{\varepsilon}$, $\chi_0(l) = \chi_0$ and we may write:

$$I(\vec{n}) = \frac{\tilde{\varepsilon}}{\chi_0}\left\{1 - exp\left[-\chi_0 L(\vec{n})\right]\right\}.$$

For highly excited noble gas states at low pressures $\chi_0 L(\vec{n})$ is smaller than unity and this relation can be simplified as:

$$I(\vec{n}) = \tilde{\varepsilon} L(\vec{n}) \tilde{\alpha}$$

and the character of the optical anisotropy is determined directly by the shape of the ionised gas entity $L(\vec{n})$. In order to obtain a clear experimental proof of self-alignment of the highly excited states of noble gases, a special U-shape discharge was used (Fig. 32). The relative efficiency of the optical and electron impact excitation is different in the specific curved region "c" with respect to the "a" and "b" regions. Because of simple geometrical considerations, the expected alignment axis under radiation self-absorption in part "c" must be the same as in regions "a" and "b". The experiment on the 645.6 nm krypton line ($5p[5/2]_3 - 6d[7/2]_4$) at 15 mTorr showed that the self-alignment signal sign was the same for "a" and "b" locations and opposite to the signal sign for the location "c". This means that self-absorption of radiation is not responsible for the self-alignment of the highly excited state . At the same time the change of sign correlates with the change of the discharge current in the chosen volumes hinting at the role of the electron impact in this case.

In order to confirm that the symmetry of the distribution of the self-alignment of highly excited atoms in the positive column is connected to the distribution of electric field in the plasma, a second model experiment was carried out. A study of the signal profile was made in the boundary region of the discharge in which the reversal of the discharge current was realised. A thorough numerical examination of the signal profile for the highly excited states of noble gases (argon and krypton) at 15 mTorr indicated that the

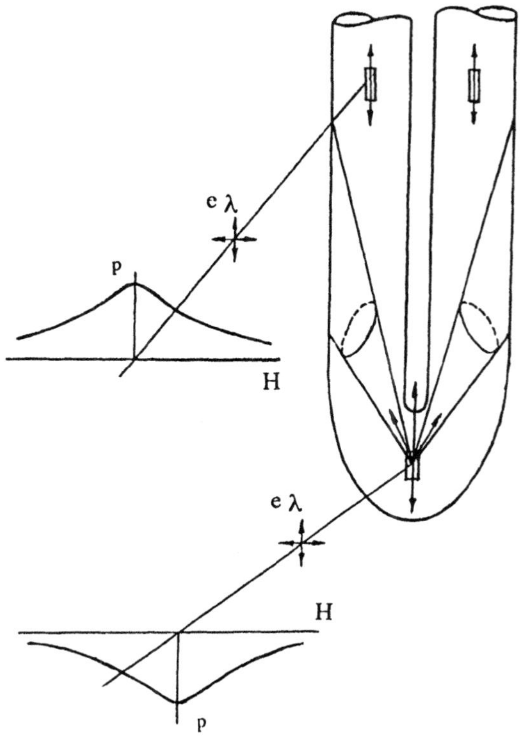

Figure 32. Self-alignment signals from different parts of a U-shaped discharge tube

dispersive part in the signal had a magnitude of approximately 5% with respect to the Lorentzian part. This was due to the small inclination of the alignment axis relative to the axis of symmetry of the discharge vessel:

$$P(H) = P_0 \left\{ \cos 2\varphi \left[1 + (2\Omega/\Gamma_2)^2\right]^{-1} - \sin 2\varphi (2\Omega/\Gamma_2)^2 \left[1 + (2\Omega/\Gamma_2)^2\right]^{-1} \right\} \tag{3.7}$$

where the inclination angle φ defines the relative contribution of Lorentzian and dispersive components.

The sign of the dispersive part was changed when the discharged current was reversed so that the angle φ was changed to $-\varphi$. This was possible only if this angle was determined by the local factor E_z/E_r ($E_{z,r}$ are respectively the axial and radial electric field components). This idea was confirmed by the observation of the signal shape at the opposite boundary region of the positive column (Fig. 33).

Taking into consideration the significance of this experiment, a special signal storage regime was implemented. For a time, the signal from the left boundary side of the discharge image was recorded and after that, by

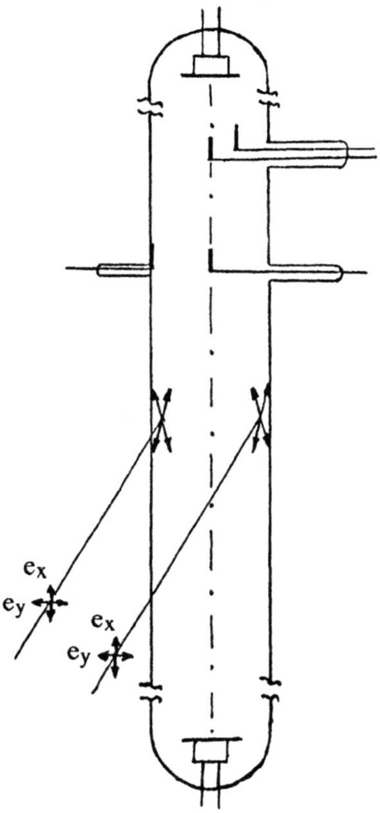

Figure 33. Scheme of an experiment for observation of self-alignment signals from the boundary regions of a d.c. discharge

shifting the projecting lens, the opposite boundary region was examined. Simultaneously the sign of the total signal $P(H)$ was changed by inverting the registration channels of the differential amplifier:

$$S' = I_y - I_x = -(I_x - I_y) = -S$$

Time averaging during the same time of the Lorentzian part of the signal brought about a considerable growth of the dispersive one, as shown in Fig. 34. This specific signal recording regime considerably suppressed the Lorentzian part of the signal. Reversing the order of the measurements gave the opposite sign of the dispersive component (Fig. 34), in accordance to the model of electron impact self-alignment. Thus, these qualitative experiments proved that the selective self-alignment of highly excited states of noble gases at low pressure is determined by electron-atom collisions.

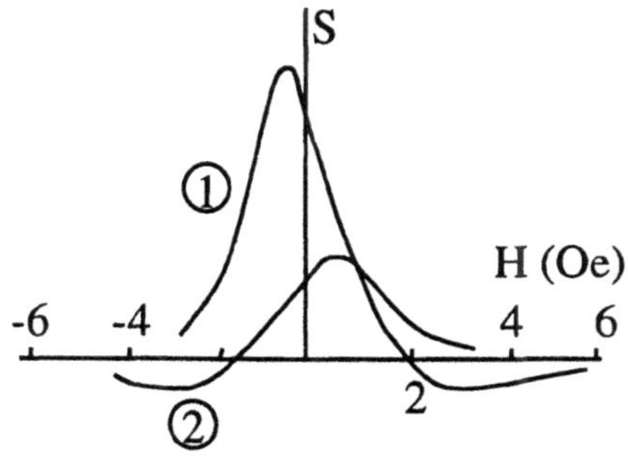

Figure 34. **Dispersive component of the self-alignment signal in 430.0 nm argon line: 1) observation starting from the left side and moving to the right side of the image 2) opposite order of observation**

3.2.2 SELF-ALIGNMENT IN A CYLINDRICAL DISCHARGE

The distribution of electron impact self-alignment in a discharge plasma must be dependent on the distribution of the internal electric field. For experimental verification, special measurements of the plasma electric field were carried out by means of the electric probe technique. One flat and two cylindrical probes were inserted into the discharge plasma (Fig. 33). The mean radial and axial potential gradients have been estimated for the same experimental conditions where the highly excited state self-alignment have been observed. For the same pressure range in all noble gases, a pronounced increase of the mean radial electric field have been detected when selective self-alignment was present. Numerical processing of the signal in a discharge tube of 22 mm showed that the inclination angle of the alignment axis $\varphi = 2^o$ was in accordance with a ratio of the local electric field projections $E_y/E_r \simeq 0.05$, and with the orientation of the local electric field in the boundary region.

The peculiarities of the self-alignment signal shape of the highly excited states of noble gases enabled to build-up the distribution of the atomic particle polarization induced by electron impact in the positive column plasma of the d.c. discharge. Experiments show that the signal shape from the central part of plasma is purely Lorentzian and the dispersive component appears only in boundary regions. Comparing these data with the distribution of self-alignment caused by radiation self-absorption in the same discharge vessel, distributions of atomic state polarisation typical for optical and electron impact processes in a plasma were derived /59/. They are presented in Fig. 35.

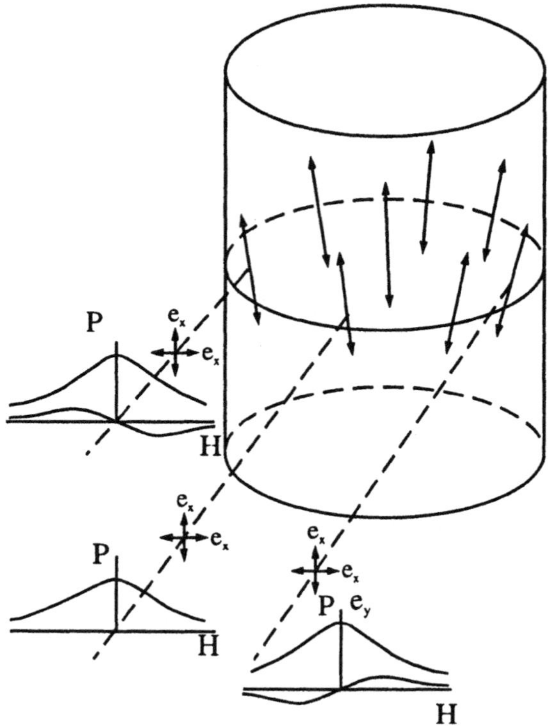

Figure 35. Distribution of self-alignment induced by electron impact in a d.c. discharge

3.3 Electric parameters in the positive column

The peculiarities of the distribution of atomic self-alignment of noble gases studied are the basis for application of the spectropolarimetric diagnostic for sensing the electric parameters of a positive column plasma. Such discharge is a very useful object, because it is possible to estimate the same quantities by using also an independent electric probe technique.

3.3.1 KINETICS OF FAST ELECTRONS

Electrons in the positive column of a plasma at low pressure may be regarded as moving inside the potential well formed by the radial potential distribution in the plasma

$$\varphi(r) = \varphi(y, r) - \varphi(0, r)$$

(r is the distance from the wall, and y is the axial coordinate) and the potential drop $\Delta\varphi$ near the wall /60/. The magnitude of this drop is much higher than the mean electron electron energy. All electrons may be divided into two sub-ensembles: electrons trapped by the wall potential drop and free fast electrons which have a nonzero probability of reaching the discharge tube wall.

In order to examine the anisotropy of the trapped electrons when the mean free-paths of electrons and ions are comparable to the discharge tube diameter and when

$$E_r = -(\partial\varphi(y,r)/\partial r) \gg E_y = -(\partial\varphi(y,r)/\partial y),$$

it is reasonable, following /60/, to introduce, for the velocity distribution function f, the following set of variables : azimuth angle ψ, local total energy $\varepsilon_0 = mv^2/2 + e\varphi(r)$, momentum $\mu = mv_\psi r$, tangent velocity projection v_ψ and dimensionless variable $\alpha = v_y\sqrt{m/2\varepsilon}$ (v_y is the axial velocity component). Then, the kinetic equation takes the form:

$$\frac{\partial f}{\partial y}\dot{y} + \frac{\partial f}{\partial r}\dot{r} + \frac{\partial f}{\partial \psi}\dot{\psi} + \frac{\partial f}{\partial \varepsilon_0}\dot{\varepsilon_0} + \frac{\partial f}{\partial \mu}\dot{\mu} + \frac{\partial f}{\partial \alpha}\dot{\alpha} = \Xi(f), \tag{3.8}$$

where $\Xi(f)$ is the collision integral.

For an asymuthally and axially symmetric distribution function, we may write:

$$\partial f/\partial y = 0,\ \partial f/\partial \psi = 0,\ \dot{\mu} = 0.$$

Expressing the derivatives $\dot{\varepsilon_0}$ and $\dot{\alpha}$ we have:

$$v_r\frac{\partial f}{\partial r} + \frac{\partial f}{\partial \varepsilon_0}ev_yE_y + \frac{\partial f}{\partial \alpha}\frac{eE_y}{\sqrt{2m\varepsilon_0}}(1-\alpha^2) = \Xi(f), \tag{3.9}$$

where the radial velocity projection v_r is expressed:

$$v_r = \pm\sqrt{\frac{2}{m}\left[\varepsilon_0(1-\alpha^2) + e\varphi(r) - \frac{\mu^2}{2mr^2}\right]}.$$

The + or - signs correspond respectively to the motion of an electron towards the wall and in the opposite direction ($\dot{r} < 0$).

The kinetic equation for electrons moving to the wall of the discharge (f_+) and to the axis (f_-) is expressed as:

$$\pm \mid v_r \mid \frac{\partial f_\pm}{\partial r} + ev_yE_y\frac{\partial f_\pm}{\partial \varepsilon_0} + \frac{eE_y}{\sqrt{2m\varepsilon_0}}(1-\alpha^2)\frac{\partial f_\pm}{\partial \alpha} = \Xi_\pm(f) \tag{3.10}$$

The distribution function for this set of variables is ambiguous. At the turning point of the radial velocity projection ($v_r = 0$), that takes place at radial distances r_{min} and r_{max}, the following condition must be fulfilled :

$$f_+(\varepsilon_0\mu\alpha, r_{min}) = f_-(\varepsilon_0\mu, \alpha, r_{min});\ \ f_+(\varepsilon_0\mu\alpha, r_{max}) = f_-(\varepsilon_0\mu, \alpha, r_{max})$$

Adding the two equations (3.10), we have:

$$| v_r | \frac{\partial \delta f}{\partial r} + ev_y E_y \frac{\partial \bar{f}}{\partial \varepsilon_0} + \frac{eE_y}{\sqrt{2m\varepsilon_0}}(1-\alpha^2)\frac{\partial \bar{f}}{\partial \alpha} = \frac{1}{2}\left[\Xi_+(f) + \Xi_-(f)\right], \quad (3.11)$$

where

$$\delta f = (f_+ - f_-)/2, \bar{f} = (f_+ + f_-)/2; \delta f(\varepsilon_0, \mu; \alpha, r_{min}) = \delta f(\varepsilon_0, \mu; \alpha, r_{max}) = 0.$$

After integration over the cross-section, we have the following equation:

$$ev_y E_y \frac{\partial f_0}{\partial \varepsilon_0} + \frac{eE_y}{\sqrt{2m\varepsilon_0}}(1-\alpha^2)\frac{\partial f_0}{\partial \alpha} = \int_{r_{min}}^{r_{max}} dr \frac{\Xi_+(f) + \Xi_-(f)}{2 | v_r |} \left(\int_{r_{min}}^{r_{max}} \frac{dr}{| v_r |}\right)^{-1} \quad (3.12)$$

where:

$$f_0(\varepsilon_0, \mu, \alpha) = \int_{r_{min}}^{r_{max}} dr \frac{\bar{f}(\varepsilon_0, \mu, \alpha)}{| v_r(\varepsilon_0, \mu, \alpha) |} \left[\int_{r_{min}}^{r_{max}} \frac{dr}{| v_r(\varepsilon_0, \mu, \alpha) |}\right]^{-1}$$

Using the weak dependence of the distribution function on the radial coordinate with an accuracy sufficient for a qualitative description, we may write

$$f_\pm(\varepsilon_0, \mu, \alpha, r) = \bar{f}(\varepsilon_0, \mu, \alpha, r) = f_0(\varepsilon_0, \mu, \alpha, r)$$

and equation (3.12) takes the form:

$$ev_y E_y \frac{\partial f_0}{\partial \varepsilon_0} + \frac{eE_y}{\sqrt{2m\varepsilon_0}}(1-\alpha^2)\frac{\partial f_0}{\partial \alpha} = \Xi'(f_0). \quad (3.13)$$

This equation has the same structure as the equation for the distribution function of electrons in an homogeneous plasma with an axial electric field; only in this case, if ε_0 is considered as the kinetic energy of electrons and α is the angle between the velocity vector $\vec{v}$ and OY axis. Using this equation, it is possible to get expressions for the multipole moments of the velocity distribution function for trapped electrons which are dependent only on the weak axial electric field E_y. It means that the motion of the trapped electrons is only sligthly anisotropic. Experimentally this is confirmed by the low degree of alignment of the deep levels of noble gases in the positive column plasma.

The axial electric field and the wall potential drop will contribute to the angular distribution of velocities of the fast free electrons which induce self-alignment of the highly excited atoms and perform direct ionization of the neutral atomic component. Anisotropy of fast electron motions is connected to the possibility to get through the wall potential barrier and

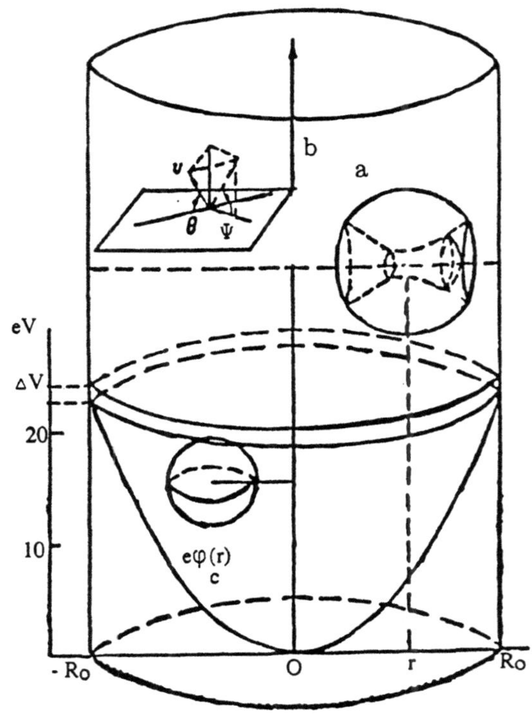

Figure 36. **Formation of loss cones in the velocity distribution of fast electrons in the positive column plasma (a), frame of reference (b), radial potential profile determined from spectropolarimetric data in 626.6 nm neon line (c)**

to the wall itself. This possibility depends on the orientation of the velocity vector.

Assuming that electrons are aborbed completely by the dielectric discharge tube wall, we may write the condition for a fast electron to reach the wall:

$$\varepsilon_0(1-\alpha^2) \geq eV_0$$

where V_0 is the wall potential. Hence, for $\varepsilon_0 > eV_0$ this condition determines the loss cone which is formed in the velocity space of the fast electrons. It is the main reason for the anisotropy of the velocity distribution of fast electrons /61/ (Fig. 36). The selectivity of the highly excited states self-alignment, and experimental estimates by electric probes of the quantity E_y/E_r at low pressures, as well as supplementary experiments described in /59/, confirm that self-alignment of the excited levels is directly determined by the formation of the loss cones . In order to calculate the shape of the loss cone in the positive column we shall use the same frame of reference with the radial vector $\vec{r}$ directed from the axis of discharge to the studied point. Within this frame of reference the electron velocity vector is characterized by the polar angles θ and ψ (Fig. 36) /62, 63/. The shape of the cone of losses is determined by the trajectory of an electron with the velocity vector $\vec{v}(\theta_0, \psi_0)$ in the point under study and tangent to the wall ($v_r = 0$, $r = R$). Using laws of conservation of energy, momentum and of the z-projection of the velocity vector:

$$m(v_y^2 + v_\psi^2(r))/2 + e\varphi(r) = \varepsilon_0 = const;$$
$$v_y = const; \quad mv_\psi r = \mu_0 = const$$

we obtain the expression describing the shape of the loss cone:

$$\sin^2\theta_0 = S_0(\varepsilon, r)/\left[\xi^2 + (1-\xi^2)\sin^2\psi_0\right] \tag{3.14}$$

where

$$S_0(\varepsilon, r) = \left[\varepsilon + e\varphi(r) - eV_0\right]/\varepsilon; \quad \varepsilon = mv^2/2 = \varepsilon_0 - e\varphi(r); \quad \xi = r/R.$$

On the discharge axis this relation simplifies as:

$$\sin^2\theta_0 \sin\psi_0 = S_0(\varepsilon, 0)$$

and corresponds to a cone axially symmetric with respect to OY axis. The distribution function for the velocity vectors out of this cone is zero. In the boundary region near the wall $\xi \to 1$ we have

$$\sin^2\theta_0 = S_0(\varepsilon, r).$$

This relation determines the loss cone with symmetry axis oriented along the $\vec{r}$ vector. In the last case the distribution function is zero for velocities inside the cone. For intermediate cases, the shape of the loss cone is much more complicated. The los cone charateristic angle, which determines the local anisotropy of the fast electrons, is dependent on the radial coordinate and finally on the local plasma potential for the given cross-section $\varphi(r)$.

3.3.2 POLARIZATION PROFILE ACROSS THE DISCHARGE IMAGE

Using (2.24), (1.44) and (1.26), the distribution of the degree of polarization across the image of a cylindrical positive column for an optically thin plasma characterized by a low degree of self-alignment may be expressed as:

$$P(x) = \frac{1}{5}w(JJ_1)\frac{\Gamma_0}{\Gamma_2}\int_{\varepsilon_t}^{\infty} d\varepsilon\,\varepsilon\sigma^{(2)}(\varepsilon)P_1(\varepsilon, x)I_1(\varepsilon, x)/\int_{\varepsilon_t}^{\infty} d\varepsilon\,\varepsilon\sigma^{0}(\varepsilon)I_1(\varepsilon, x)$$

$$P_1(\varepsilon, x) = I_1^{-1}(\varepsilon, x) \int_r^R \frac{dr\, r}{\sqrt{r^2 - x^2}}$$

$$\left[\frac{1}{2}\left(2 - \frac{x^2}{r^2}\right) {}^{\vec{r}}f_{2+}^{(2)}(\varepsilon, r) + \sqrt{\frac{3}{2}\frac{x^2}{r^2}}\, {}^{\vec{r}}f_0^{(2)}(\varepsilon, r)\right]$$

$$I_1(\varepsilon, x) = \int_x^R \frac{dr\, r}{\sqrt{r^2 - x^2}} f_0^{(0)}(\varepsilon, r) \qquad (3.15)$$

According to (1.41) and (3.14), and taking into account loss cones, the multipole moments of the fast electron distribution function take the form:

$$f_0^{(0)}(\varepsilon, r = \frac{2}{\pi} f_0(\varepsilon_0) \int_a^1 \frac{d\tau}{\sqrt{1 - \tau^2}} \sqrt{1 - B(\varepsilon, r)}$$

$${}^{\vec{r}}f_0^{(2)}(\varepsilon, r) = -\frac{5}{\pi} f_0(\varepsilon_0) \int_a^1 \frac{d\tau}{\sqrt{1 - \tau^2}} B(\varepsilon, r) \sqrt{1 - B(\varepsilon, r)}$$

$${}^{\vec{r}}f_{2+}^{(2)} = \sqrt{\frac{2}{3}}\frac{5}{\pi} f_0(\varepsilon) \int_a^1 \frac{d\tau\,(1 - 2\tau^2)}{\sqrt{1 - \tau^2}} \left[2 + B(\varepsilon, r)\right] \sqrt{1 - B(\varepsilon, r)} \qquad (3.16)$$

where $a = 0$ for $\xi > S_0(\varepsilon, r)$, $a = \left[(S_0 - \xi^2)/(1 - \xi^2)\right]^{1/2}$ for $\xi < S_0(\varepsilon, r)$, $\tau = \sin\psi$, $B(\varepsilon, r) = S_0(\varepsilon, r)/\left[\xi^2 - (1 - \xi^2)\tau^2\right]$.

Model calculations not accounting for collisions have been carried in the first stage using expressions (3.15) and (3.16). The velocity distribution function of fast electrons was assumed to be Maxwellian

$$f_0(\varepsilon_0) \sim \exp\left(-\varepsilon_0/eV_e'\right),$$

where V_e' is the effective electron temperature in the energy scale, characterising the decrease of $f_0(\varepsilon_0)$ for fast electrons. The radial potential profile was approximated by a power law:

$$\varphi(r) = (V_0 - \Delta\varphi)(r/R)^n$$

where V_0 is the potential drop between the wall and the discharge axis.

For a precise quantitative solution to this problem one should know the energy dependence of the alignment cross-section near the threshold of excitation of the level under study. The behaviour of the impact alignment cross-section may be very complicated. Reliable information for the alignment cross-section is known only for a limited number of atomic states. Nevertheless, for the qualitative analysis of the polarization profiles, the transversal distribution of polarization is well approximated by the dependence for electrons of $P_1(\varepsilon_t, x)$ on the threshold energy ε_t. Indeed, due to the rapid fall of the distribution function of fast electrons, the quantity

$I_1(\varepsilon, x)$ from (3.15) also decreases exponentially for high ε. Therefore the energy dependencies of the products $\sigma^{(2)}(\varepsilon)I_1(\varepsilon, x)$ and $\sigma^{(0)}(\varepsilon)I_1(\varepsilon, x)$ are localized around the threshold energy. It means that the real range of integration in (3.15) is the interval between ε_t and ε_t plus some units of eV_e'. The last value for the effective energy excess over the threshold of inelastic electron-atom interaction is lower than the mean electron temperature /60/. So the range of integration does not exceed some electron-volts. For such approximation we may write the expression (3.15) in the simplified form:

$$P(x) = (\Gamma_0/5\Gamma_2)\, w(JJ_1)\sigma^{(2)}(\varepsilon')P_1(\varepsilon', x)I_1(\varepsilon', x)\varepsilon'/\sigma^{(0)}(\varepsilon'')I_1(\varepsilon'', x)\varepsilon'',$$

$$P(x) \cong C_J P_1(\varepsilon_t, x) \qquad \text{if} \qquad \varepsilon' \approx \varepsilon'' \approx \varepsilon_t \tag{3.17}$$

where

$$C_j = (\Gamma_0/5\Gamma_2)w(JJ_1)\sigma^{(2)}(\varepsilon')/\sigma^{(0)}(\varepsilon'').$$

At threshold energy, C_j may be estimated, using (1.58a), as

$$C_j = (\Gamma_0/5\Gamma_2)\sqrt{2J+1}w(JJ_1)\alpha(S).$$

Calculations of $P_1(\varepsilon, x)$ have been made within the most interesting range of energies close to V_0 and the excitation potentials of the atomic states studied (Fig. 37). For an arbitrary radial dependence of the potential, a weak sensitivity to V_e' is obtained. Despite the variation of V_e' within wide range $(0.05 - 0.2)V_0$, no visible difference of $P_1(\varepsilon, x)$ was observed. At the same time these calculations demonstrate a rather high sensitivity of the spectropolarimetric effects to the radial dependence of potential in the plasma. Fig. 37 shows the results of calculations for a parabolic and for power law approximations $(n > 2)$ of $\varphi(r)$. For $\varepsilon < eV_0 - e\varphi(r)$, electrons in the point under study have no possibility to go to the wall, cutting through the potential barrier. Therefore, in the velocity distribution of such electrons there are no loss cones. This is reflected in the presence of a nonsensitive flat region for the distribution $P_1(\varepsilon, x)$ ($P_1(\varepsilon, x) = 0$ for $\varepsilon < e\Delta\varphi$ in all discharge volume). This case corresponds to an experimentally detected zero self-alignment degree for the deep $2p$ lecel of noble gas atoms at low pressures.

For $e\Delta\varphi < \varepsilon < eV_0$, there are regions in the discharge where loss cones are formed, but these regions are localized near the wall, where $e\varphi > eV_0 - \varepsilon$. The polarization anomalies near the wall are explained by this effect. The small polarization in the center of the image is determined only by a volume effect, that is the contribution of the boundary wall regions positioned on the line of sight. Finally for $\varepsilon > eV_0$ the loss cone exists for all points

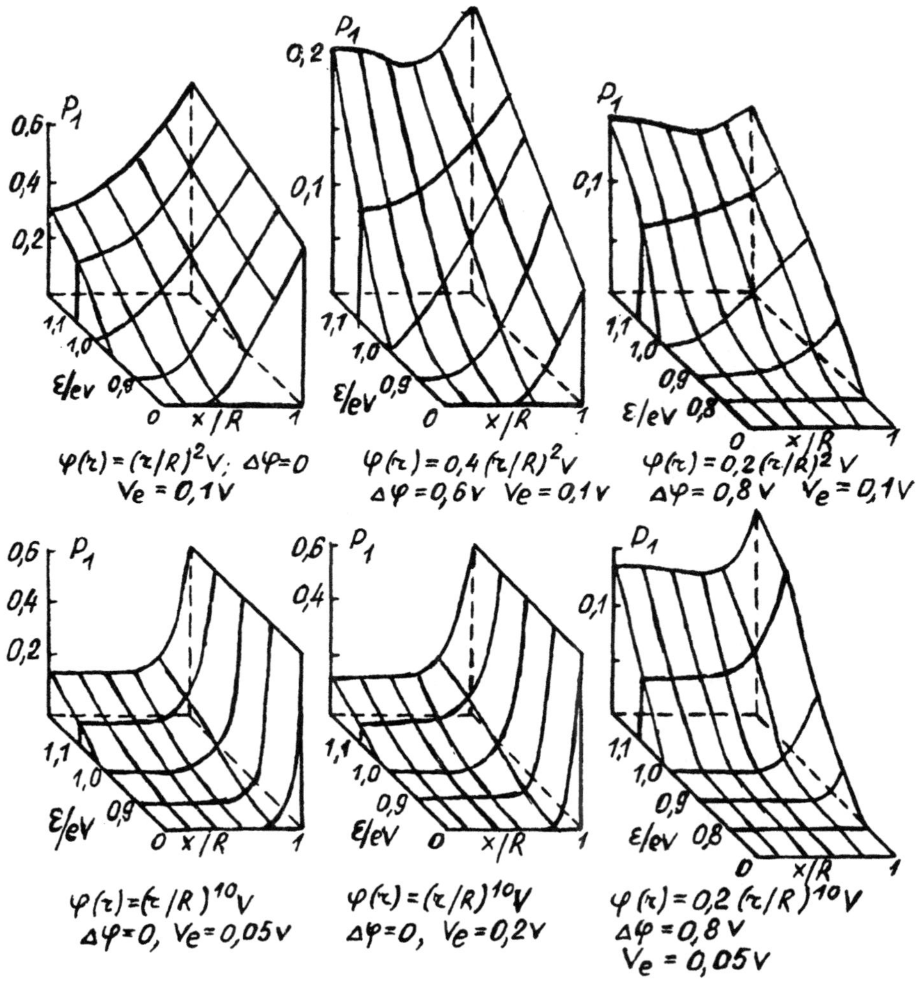

Figure 37. Calculated polarization energy and reduced radial distance dependencies for different sets of parameters in a cylindrical d.c. discharge plasma (with no account for elastic electron-atom collisions)

of the discharge. This is reflected in the strong polarization of emission for all points of the discharge image, which is growing with ε. This effect corresponds to the increase of the characteristic angle of the loss cone.

The polarization profile $P_1(\varepsilon, x)$ for $\varepsilon > eV_0$ reflects the radial variation of potential $\varphi(r)$ quite well. This can be easily explained for a ionized gas confined within parallel dielectric walls. In this case the symmetry axis of the loss cone is oriented along the OX direction, perpendicular to the wall, and the characteristic angle of the loss cone, θ_0, is determined by the relation $\cos^2\theta_0 = [eV_0 - e\varphi(x)]/\varepsilon$, where V_0 is the potential drop between the wall and the ionized gas and $\varphi(x)$ is the coordinate potential distribution. For $\varepsilon > eV_0$, the quadruple and the zero-order multipole moments of the velocity distribution function take the form:

$$f_0^{(2)}(\varepsilon, x) = (5/2) f_0(\varepsilon) \cos\theta_0(\cos^2\theta_0 - 1)$$
$$f_0^{(0)}(\varepsilon, x) = f_0(\varepsilon) \cos\theta_0$$

The degree of polarization for the threshold excitation according to (3.17) will be given by:

$$P(x) = C_J f_0^{(2)}(\varepsilon_t, x)/f_0^{(0)}(\varepsilon_t, x) = C_J(5/2)\left[eV_0 - e\varphi(x) - \varepsilon_t\right]/\varepsilon_t \quad (3.18)$$

It means that in the plane parallel case P(x) is directly determined by the potential profile $\varphi(x)$. In a cylindrical discharge tube, the $P_1(\varepsilon_t, x)$ distribution is mostly close to the $\varphi(r)$ function when $\Delta\varphi \approx V$ (Fig. 37). In this case, the shape of the loss cone is similar to that for a plane parallel layer. When $\Delta\varphi$ decreases, the shape of the loss cone becomes dependent on the curvature of the wall and the $P_1(\varepsilon, x)$ distribution differs from $\varphi(r)$. However, the results of numerical calculations indicate that the transversal distribution of the degree of polarization of the positive column may be used for the determination of the potential profile $\varphi(r)$.

By a qualitative examination of the polarization profiles in the positive column of a low pressure gas discharge, assuming that the velocity distribution of the fast electrons where the loss cones are formed is stationary and not dependent on the axial electric field, the Influence of electron-atom collisions may be estimated using Liouville-Vlasov principle.

Using the total electron energy ε_0 and the distance along the electron trajectory l as variables, and taking into consideration that $\dot{\varepsilon_0} = 0$ along the electron trajectory, we may write the following kinetic equation:

$$\sqrt{2\left[\varepsilon_0 - e\varphi(r)\right]/m}\,\frac{\partial f(\varepsilon_0, l)}{\partial l} = \Xi(f) \quad (3.19)$$

The collision integral for a low temperature gas discharge plasma, with account for electron-atom collisions /29/, may be represented as:

$$\Xi[f(\varepsilon, \vec{n}] = \Phi[f(\varepsilon)] - \nu(\varepsilon)\left[\overline{f(\varepsilon, \vec{n})} - f(\varepsilon, \vec{n})\right] + \Xi^*[f(\varepsilon, \vec{n})] \quad (3.20)$$

where $\Phi[f(\varepsilon)]$ is the Focker-Planck term describing the electron energy drift resulting from elastic electron-atom collisions , $\nu(\varepsilon)$ is the frequency of elastic electron-atom collisions, $\vec{n'} = \vec{n}/v$, $\overline{f(\varepsilon,\vec{n})}$ is the distribution function in the point under study averaged over all directions:

$$\overline{f(\varepsilon,\vec{n})} = \int d\vec{n'}\, f(\varepsilon,\vec{n'})\sigma_e(\varepsilon,\vec{n},\vec{n'}) / \int d\vec{n'}\, \sigma_e(\varepsilon,\vec{n},\vec{n'}) \tag{3.21}$$

where $\sigma_e(\varepsilon,\vec{n},\vec{n'})$ is the differential cross-section of elastic electron-atom scattering.

In the simplest case, σ_e is not angular dependent and $\overline{f(\varepsilon,\vec{n})}$ represents the isotropic part of the distribution function $\overline{f(\varepsilon,\vec{n})} = f_0^{(0)}(\varepsilon)$, and $\Xi^*(f)$ is the contribution to the collision integral due to the elastic electron-atom interaction.

For the study of the anisotropy of the fast electrons in the tail of the velocity distribution, which is specific for spectropolarimetric measurements, the Fokker-Planck term as well as inelastic collisions may be neglected and we have the simplified equation:

$$\frac{\partial f(\varepsilon_0,l)}{\partial l} = \left[\overline{f(\varepsilon_0,l)} - f(\varepsilon_0,l)\right]/\lambda_e(l) \tag{3.22}$$

where $\lambda_e(l) = v(l)/\nu(l)$ is the mean free path of electrons.

Therefore, in order to estimate the velocity distribution function in an arbitrary point of plasma for a given potential profile $\varphi(r)$, it is necessary to determine its value in the other points of the discharge. As the first approximation, this iteration procedure may be realized using the equilibrium isotropic distribution function $f_0(\varepsilon_0)$ for the center of the discharge. The solution of equation (3.22) takes the form:

$$f(\varepsilon_0,l) = f(\varepsilon_0,0)\exp\left[-\int_0^l \frac{dl'}{\lambda_e(l')}\right] + \int_0^1 \exp\left[-\int_0^1 \frac{dl''}{\lambda_e(l'')}\right]\frac{f_0(\varepsilon_0)}{\lambda_e(l'')}dl'' \tag{3.23}$$

Here, $f(\varepsilon_0,0)$ is the distribution function for the initial point of trajectory at $l=0$. For the trajectories originating at the discharge tube wall $f(\varepsilon_0,0) = 0$. Within the energy range of the electrons responsible for the emission spectrum of noble gases, the mean free path $\lambda_e = \text{const}$ and the expression (3.23) is easily transformed to:

$$f(\varepsilon_0,l) = f_0(\varepsilon_0) + [f(\varepsilon_0,0) - f_0(\varepsilon_0)]\exp(-l/\lambda_e). \tag{3.24}$$

This result may be interpreted as the exponential variation of the distribution function moving along the electron trajectory to its mean and isotropic value with the characteristic length λ_e. For all electron trajectories crossing, the discharge wall $f(\varepsilon_0,l) = 0$, if $\lambda \gg R$ and for remaining trajectories $f(\varepsilon_0,l) = f_0(\varepsilon_0)$.

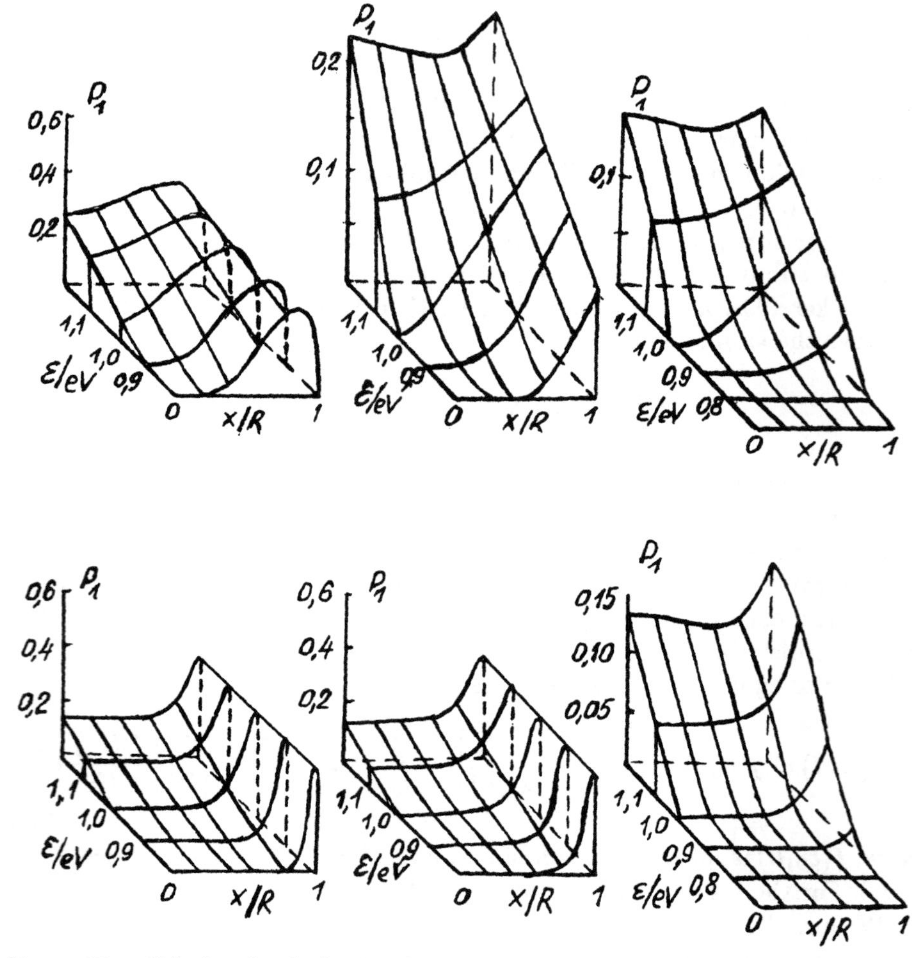

Figure 38. **Calculated polarization energy and reduced radial distance dependencies for different sets of parameters in a cylindrical d.c. discharge plasma (with account for elastic electron-atom collisions for a electron mean-free-path of 6 R)**

Accounting for collisions, a qualitative analysis of the polarization profiles is made by adding the small factor

$$[f_0(\varepsilon + \varphi(r)e) - f_0^{(0)}(\varepsilon, r)] \cdot [1 - \exp(-2R/\lambda_e)]$$

to the distribution function, $f_0^{(0)}(\varepsilon, r)$, in all points of the plasma. This transformation describes the partial filling of the loss cones by the elastically scattered electrons with an isotropic distribution function. For different λ_e, the results of calculations are presented at Fig. 38. For a high potential drop $\Delta\varphi \approx V_0$, accounting for collisions does not significantly influence the polarization profiles. For $\Delta\varphi \ll V_0$ the polarization maxima are formed near the wall. These maxima may be easily explained within the framework of the simple plane-parallel plasma model. Taking into consideration the collisional filling of the loss cones in a plasma maintained in an infinite plane parallel layer of width d, we may write for the zero-order and quadrupole moments of the velocity distribution function :

$$f_0^{(2)}(\varepsilon, x) = (5/2)\gamma f_0(\varepsilon_0)\cos\theta_0(\cos^2\theta_0 - 1)$$
$$f_0^{(0)}(\varepsilon, x) = f_0(\varepsilon)\left[1 - \gamma(1 - \cos\theta_0)\right]$$

where

$$\cos^2\theta_0 = [eV_0 - e\varphi(x)]/\varepsilon,\ \gamma = \exp\left(-d/\lambda_e\right).$$

According to (3.17), for the degree of polarization of the studied spectral line at the threshold conditions we have:

$$P(x) = C_J(5/2)\gamma\cos\theta_0(\cos^2\theta_0 - 1)\left[1 - \gamma(1 - \cos\theta_0)\right]^{-1} \tag{3.25}$$

This function describes the polarization profile in the vicinity of the boundary where it is possible to neglect the curvature of the wall. For $0 < \gamma < 1$, when the characteristic angle θ_0 of the cone is increased in the range $[0, \pi/2]$, the $P(x)$ dependence has a maximum, and when $\gamma \to 0$ (complete collisional filling) this maximum will be at the point x_0, determined by the relationship $e\varphi(x_0) = \varepsilon_t/3 - eV_0$. For negligible electron-atom collisions $(\gamma = 1)$, $P(x)$ is represented by a monotonous curve with no maximum near the wall. These peculiarities of the plane parallel layer qualitatively represent the boundary behaviour of $P_1(\varepsilon, x)$ in a cylindrical discharge tube (Fig. 38).

3.3.3 RADIAL ELECTRIC FIELD

The solution of spectropolarimetric diagnostic problems may be realized either applying general methods of solving an ill-posed inverse spectroscopic problem /45/ or using the technique of numerically simulating the experiment. This technique is based on the repeated numerical solution of the direct problem for a variety of parameters. For the determination of the electric field in a cylindrical positive column, we shall use the integral relation (3.15) for the polarization profile. In this case, the complex expression for the multipole moments of the velocity distribution function is constructed by integration over θ and ψ of the isotropic function, $f_0(\varepsilon_0)$, outside the loss cone, and inside the loss cones of the functions $f_0'(\varepsilon)$ and $f_0''(\varepsilon)$ resulting from the elastic scattering into the cones facing to the nearest and opposite walls of the discharge tube (Fig. 36). Within this approximation, the expression (3.16) is transformed to:

$$f_0^{(0)}(\varepsilon,r) = \frac{2}{\pi} f_0(\varepsilon_0) \left[\gamma \int_a^1 \frac{d\tau \sqrt{1 - B(\varepsilon,r)}}{\sqrt{1-\tau^2}} + \frac{\pi}{2}(1-\gamma) \right]$$

$$^{\vec{r}}f_0^{(2)}(\varepsilon,r) = -\frac{5}{\pi}\gamma f_0(\varepsilon_0) \int_a^1 \frac{d\tau\, B(\varepsilon,r)}{\sqrt{1-\tau^2}} \sqrt{1 - B(\varepsilon,r)}$$

$$^{\vec{r}}f_{2+}^{(2)}(\varepsilon,r) = \sqrt{\frac{2}{3}}\frac{5}{\pi}\gamma f_0(\varepsilon_0) \int_0^1 \frac{d\tau\,(1-2\tau^2)}{\sqrt{1-\tau^2}} \left[2 + B(\varepsilon,r)\right] \sqrt{1 - B(\varepsilon,r)} \tag{3.26}$$

where

$$\gamma = (1/2)[2f_0(\varepsilon_0) - f_0'(\varepsilon) - f_0'']/f_0(\varepsilon_0)$$

is the coefficient accounting for the collisional filling of the loss cones, and $f_0'(\varepsilon_0)$, $f_0''(\varepsilon)$ are the additional factors resulting from the collisional filling of the loss cones facing the opposite walls of the discharge tube. Assuming that the frequency of elastic electron-atom collisions for noble gases considerably exceeds that of inelastic collisions, it is possible to estimate $f_0'(\varepsilon)$ and $f_0''(\varepsilon)$ according to (3.24) as follows:

$$f_0'(\varepsilon) = f_0(\varepsilon_0)\left[1 - \exp\left(-\frac{R+r}{\lambda_{et}}\right)\right]; \quad f_0'' = f_0(\varepsilon_0)\left[1 - \exp\left(-\frac{R-r}{\lambda_{et}}\right)\right]$$

Here λ_{et} is the mean free-path for the threshold electrons performing the direct excitation of atoms.

The expression for γ then takes the form:

$$\gamma = \gamma(r) = \frac{1}{2}\left\{\exp\left[-\frac{R}{\lambda_{et}}(1+\xi)\right] + \exp\left[-\frac{R}{\lambda_{et}}(1-\xi)\right]\right\} \tag{3.27}$$

Assuming a parabolic radial dependence of the potential in the central part of the plasma and a small relative contribution of the electrons falling to the wall, the distribution function for fast electrons, $f_0(\varepsilon_0)$, may be expressed as /60/:

$$f_0(\varepsilon_0) = [(\varepsilon_0/\varepsilon_x') - 1]^{1/2} \mathcal{K}_{1/(\mathrm{p}+2)} \left\{ [(\varepsilon_0 - \varepsilon')/\mathrm{eV}_\mathrm{e}']^{\frac{p}{2}+1} \right\}$$

where $\mathcal{K}_{1/(\mathrm{p}+2)}$ is the MacDonald function , ε' is the first excitation potential of atoms, p is a parameter indicating the level of approximation of the cross section of excitation, and eV_e' is the characteristic temperature describing the decrease of the distribution function in the rapid part. This parameter is dependent on the longitudinal electric field in the plasma which was experimentally estimated by the probe method.

Numerical calculations of polarization profiles have been performed /62/, with an electron impact alignment cross-section calculated by the formulae (1.59), using either the differential cross-section or the threshold approximation (3.17). The last approximation is justified firstly by the rapid fall of $f_0(\varepsilon_0)$ for $\varepsilon_0 > \varepsilon'$ and secondly by the predominant contribution to excitation and alignment of electrons at threshold energy. The radial dependence of the potential profile is approximated by the function

$$\varphi(r) = (V_0 - \Delta\varphi)(r/R)^n$$

inside the bulk of plasma and the sharp potential drop within the wall sheath. Parameters V_0, $\Delta\varphi$, n, V_e' and the energy dependence of the tail of the distribution function have been varied in the process of calculation. Electron-atom collisions have been accounted for in two ways: constant independent collisional filling of the loss cones (γ = const) and radially dependent collisional filling according to expression (3.27). Calculations that do not take into account collisions have also been made.

In analysing the informative possibilities of the spectropolarimetric diagnostic for the positive column plasma of d.c. discharge, the simulating calculations have firstly been implemented for the different characteristic parameters of the problem (Fig. 39). Variations of γ, V_0, $\Delta\varphi$, V_e' and n parameters in the range close to the experimental values showed that for $0 < x < R$ the polarization profile is represented by a monotonous function having in some cases one maximum. The degree of polarization showed to be strongly dependent on the γ parameter. Accounting for the collisional filling of the loss cones ($\lambda_e/R \gg 1$, $\gamma = 1$) and for the threshold excitation, the degree of polarization near the wall is rather high. Accounting for the collisional filling of the cones of losses by using the radial dependent $\gamma(r)$ parameter (3.27) ($\lambda_e/R \simeq 0.33$ for experimental conditions) leads to the decrease of the degree of polarization to zero at the distance λ_e from the wall. It reflects the isotropisation of the velocity distribution function of fast electrons (Fig. 39). Calculations with the constant coefficient $\gamma = 0.1$, that means a 90% filling of the loss cone by elastically scattered electrons, showed that in this case the degree of polarization is decreasing much faster for $x \to R$.

For $R/\lambda_e > 0.1$, the elastic electron-atom collisions of fast electrons significantly influence the polarization profile; in the opposite case ($R/\lambda_e <$ 0.1), neglecting collisions introduces errors not exceeding 10 - 15%. For $R/\lambda_e \geq 1$, correctly accounting for electron-atom collisions is only possible

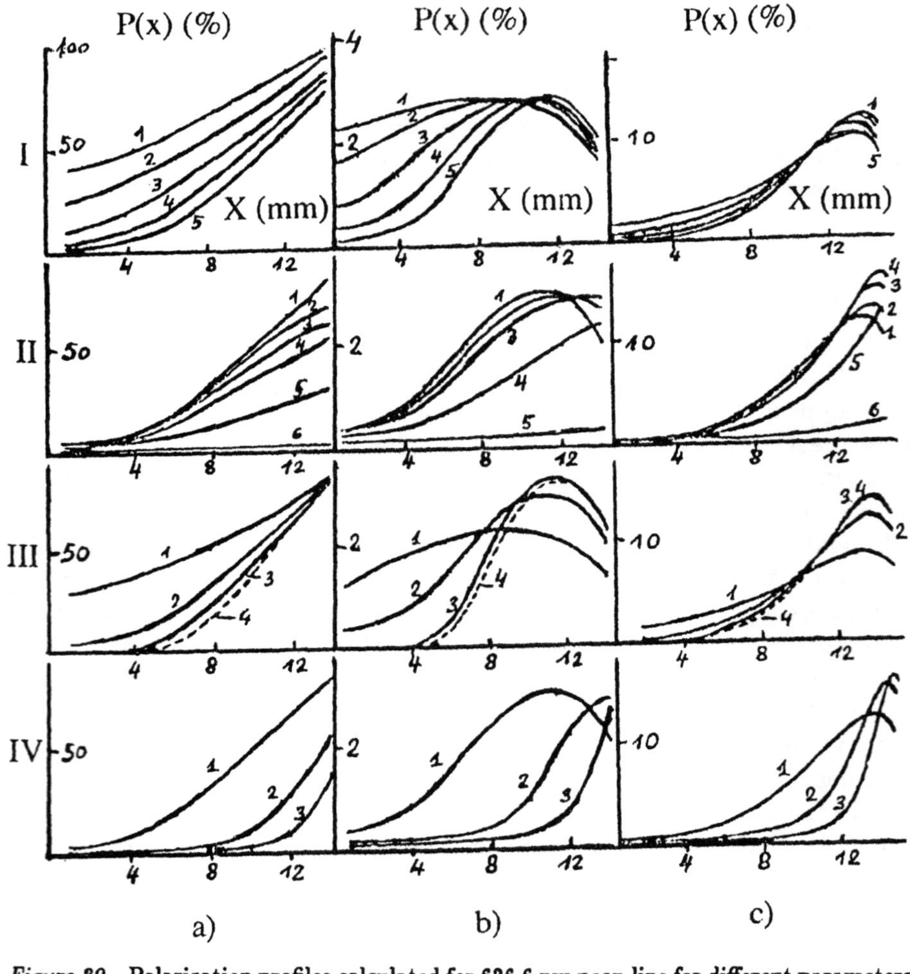

Figure 39. Polarization profiles calculated for 626.6 nm neon line for different parameters of a d.c.discharge: Vo=24 eV, $\Delta\varphi = 0$, Ve' = 3 eV, n=2: I) curves 1-5: Vo=18, 20 eV, II) curves 1-6: $\Delta\varphi = 0.1$, 2.4, 4.8, III) curves 1-3: Ve'= 10,3,1 eV (curve 4 corresponds to threshold electron impact alignment cross section), IV) curves 1-3: n= 2, 5, 10 (a, b, c correspond respectively to a mean-free-path $\lambda = 0$, 0.1 R and to $\lambda(r)$ as defined in the text)

by taking into consideration the shape of the discharge volume. In the case of $\gamma \neq 1$ and of a small potential drop, $\Delta\varphi$, a maximum of the polarization takes place (see Fig. 39a, b,c) which is easily explained in the plane parallel plasma layer model.

The quantity V_0 influences the characteristic angle of the loss cone and thus the $f_q^{(2)}(\varepsilon)$ and $P(x)$ functions in the axial part of the plasma. Increasing the effective temperature V_e' for a given V_0 broadens the maximum of the polarization profile, which reflects the increase of the relative part of the free electrons in the axial region and as a result the increase of the degree of polarization in this part. Comparison of the calculated profiles $P(x)$ for different $f_0(\varepsilon_0)$ dependencies for the tail of the distribution function (Maxwell or McDonald with the same temperature) shows that with reasonable accuracy (less than 1%) the energy dependence of the tail of $f_0(\varepsilon_0)$ does not influence the polarization profile. The increase of the exponent n in the radial profile of the potential $\varphi(r)$ causes a drift of the polarization maximum to the wall (Fig. 39), which is easily explained within the loss cone model.

The collisional filling of the loss cones by electrons, elastically scattered for the experimental conditions ($\lambda_e = 5$ mm), leads to a shift of the polarization maximum to the wall according to the model. At a distance λ_e from the wall, collisional isotropization of the distribution function takes place. Polarization profiles calculated for threshold excitation (3.17) (dashed curves on Fig. 39) did not differ considerably from that for excitation at low temperature ($eV_e' = 1$ eV). Such behaviour indicates a relatively weak energy dependence of $\sigma^{(2)}(\varepsilon)$ at energies higher than the threshold and confirms the role of threshold electrons in impact self-alignment for low temperature plasmas. Independent experimental support of this result is the selectivity of electron impact self-alignment observed in the positive column plasma /55, 61/.

Polarization profiles are strongly dependent on the radial dependency of the potential $\varphi(r)$, which is determined by the set of parameters V_0, $\Delta\varphi$, n. The cross-correlation between these parameters is relatively low. The quantity $\Delta\varphi$ influences the polarization profile $P(x)$ and the height of the wall polarization maximum. The exponent n in the $\varphi(r)$ profile determines the position of this polarization maximum with respect to the wall and V_0 influences the polarization dependence $P(x)$ for the axial part of the discharge image. The near-wall part of the polarization profile is the most sensitive to the parameters $\Delta\varphi$ and n in the case of a relatively sharp change of the potential near the wall (Fig. 39).

Reliability of the spectropolarimetric determination of the electric parameters in the positive column may be enhanced significantly using the supplementary experimental information about the plasma parameters which influence the polarization profile: λ_e, V_e, V_e'. Experimental estimates by the electric probe method show that, for the present experimental conditions,

V_e does not exceed some V cm^{-1}. Using /60/ and the low sensitivity of the $P(x)$ dependence to V_e' (Fig. 39), it is possible to use a rough estimate of V_e' in order to determine the electric parameters by this technique.

The second step of numerically simulating the experiment is connected to the estimate of the stability of the spectropolarimetric diagnostic and the assessment of the accuracy of this technique. For this purpose, the polarization profile $P(x)$ for the set of parameters $V_0 = 24eV$, $\Delta\varphi = 2.4eV$, $n = 2$, $V_e' = 3eV$, $\gamma = 0.1$ were calculated. Then, the experimental errors were simulated by adding a random quantity δ with a normal statistical distribution and variable dispersion. Using this artificial polarization profile $P_{exp}(x)$ with an introduced simulated experimental error and performing the standard least square fit technique by minimizing the function

$$\Phi_c = \sum_i [P_{exp}(x_i) - P(x_i)]^2$$

where i is the reference number of the point on the polarization profile, the principal electric parameters of the potential curve V_0, $\Delta\varphi$ can be determined. Calculations showed that for a satisfactory signal to noise ratio ($\delta < 10\%$), the minimization of Φ_c function is attained for the same set of parameters of the problem as the initial one. The growth of the experimental error ($\delta = 10, 20, 30\%$) leads to an increase of the errors on the determined parameters. The relative error on V_0 does not exceed 4% but the relative error on $\Delta\varphi$ is much higher (50, 100, 200% respectively). Therefore data derived from the numerical simulation experiment show that to an experimental accuracy of 10% it is possible to determine the parameter V_0 and to have quite a reliable estimate for $\Delta\varphi$.

The spectropolarimetric determination of the potential profile in the positive colunm plasma have been performed for a neon discharge at 0.2 Torr and 25 mA. Using the measurement of the longitudinal electric field (1.8 eV), we calculated the parameter V_e' by the formula given in /60/ ($V_e' =$ 2.6 eV). The experimental polarization profile on the 626.6 nm neon line was used ($\Gamma_2/\Gamma_0 = 1.1$). Taking into consideration the low cross-correlation of V_0 and n parameters, the minimization of the function Φ_c was made in two steps. At the first step, for $n = 2$ the following set of parameters were determined: $\gamma = 0.11(1)$, $V_0 = 24(1)eV$, $\Delta\varphi = 1.0(5)eV$. At the second step, γ and V_0 were fixed and the two parameters $\Delta\varphi$ and n were obtained ($\Delta\varphi = 1.0(5)$eV, $n = 1.96(4)$). For this set of parameters and an accuracy of the experimental data better than 10%, the function Φ_c reached 6%. The experimental data are compared with the calculated ones in Fig. 40. The stability of the solution is confirmed by the small degree of uncertainty of the determined parameters and the high sensitivity of the calculated polarization profile to the electric parameters of the plasma.

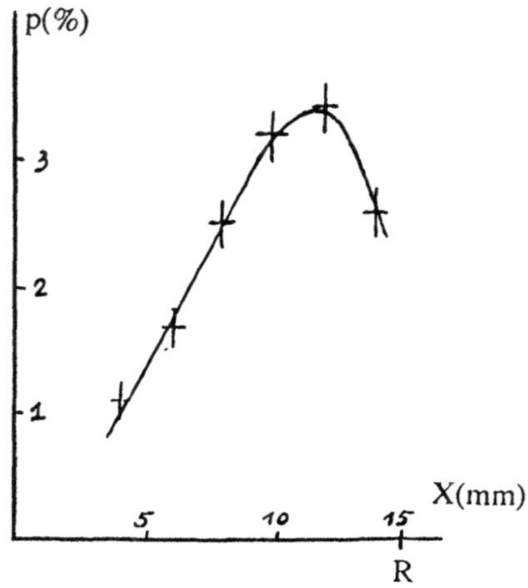

Figure 40. **Comparison of the calculated polarization profile for the 626.6 nm neon line with experimental data (crosses) for a d.c. discharge (R = 15 mm, p = 0.2 Torr, i = 25 mA)**

As a result, the radial profile of the potential in the positive column discharge plasma was found to be $\varphi(r) = 23(r/R)V$. Independent electric probe test experiments of the potential drop between two points inside the plasma confirmed the results of spectropolarimetric measurements (Fig. 36). Therefore, spectropolarimetric measurements of radiation coming from the positive column of the d.c. discharge showed their usefulness for the remote sensing of the electric field within an ionized gas.

CHAPTER 4

POLARIMETRIC SENSING OF A HIGH FREQUENCY DISCHARGE

High frequency discharges and specially capacitive discharges represent laboratory plasma entities for which the spectropolarimetric characteristics of emission are very well pronounced. The specific structural nature of the capacitive discharge, if the frequency of the alternative field is lower than the electron Langmuir frequency, is determined by the localization of practically all alternative electric field within the thin electrode sheath. Only a small part of the high frequency field penetrates to the central parts of the plasma. These features of the discharge are reflected in its different local electron kinetics. The spectropolarimetric method described in this chapter appears to be a unique tool for the remote sensing of the structural plasma parameteres such as energy delivery and the internal electric field strength.

4.1 Self-alignment generation

In spite of a long history of experimentation, the physical processes in the high frequency discharge still remain a subject of detailed interest. The interest in this discharge is determined by its wide applications as a light source with high spectral efficiency for the emission lines of the majority of elements, in quantum standards, quantum magnetometers, spectral analysis and as an active medium of gas lasers. Spectral light sources on the basis of the high frequency discharge are spherical or cylindric lamps filled with a buffer gas and the studied element, which are excited by a high frequency oscillator. Discharge is induced within either the inductive coil or capacitor and by the way of excitation high frequency discharges are divided into two forms: inductive H-discharge and capacitive E-discharge.

4.1.1 RADIATION SELF-ABSORPTION

Self-alignment of the noble gas atoms in H-discharge was studied in the spherical lamp of 2.8 cm diameter fixed inside the solenoid of a high frequency oscillator operating at 40 MHz. The axis of the solenoid was parallel to the direction of observation and the vector of the magnetic field strength of the magnetic spectropolarimeter. E-discharge was excited at 100 MHz in different lamps inside the plane capacitor with the plates oriented parallel to the magnetic field /64, 65/.

Self-alignment on H-discharge was observed on different neon lines corresponding to transitions between 1s and 2p configurations (640.2, 613.4, 630.5, 692.9, 626.6, 671.7, 603.0, 616.4, 656.5 nm) at the pressures 0.1-2

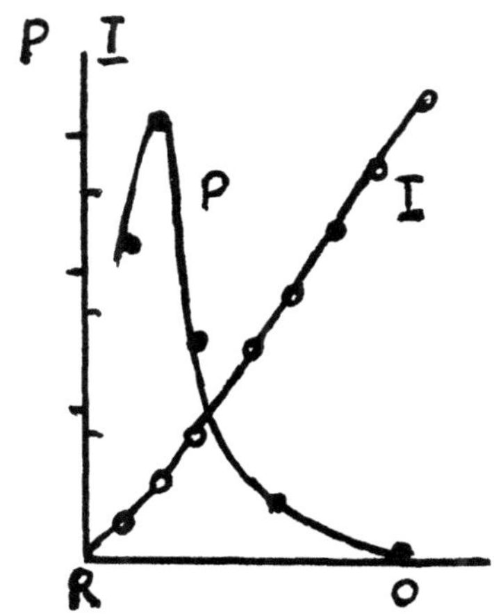

Figure 41. Radial dependence of the first (I) and second (P) Stokes parameters for the line 626.6 nm in a spherical high frequency inductive discharge at a 40 MHz frequency and at pressure 1 Torr

Torr. In E-discharge the self-alignment of the same 2p states was detected for the wider range of pressures 0.003-7 Torr.

The first observations of spectropolarimetric effects in different locations of the discharge indicated the complex nature of this phenomenon. Scanning the detected light beam across the image with the help of the square diaphragm mounted in the plane of the entrance slit of the monochromator, it was shown that the degree of polarization of all spectral lines decreased as the diaphragm approached the center of the image of H-discharge. The intensity of radiation increased at the same time (Fig. 41). For the experimental conditions the relationship $\chi_0 R \approx 1$ was fulfilled for all the spectral lines studied. It means that radial symmetry of the resonance radiation took place and the degree of the optical field anisotropy was increasing in the boundary parts of the discharge. The correlation of the degree of polarization in the outer parts of H-discharge with the optical field anisotropy enables one to explain the mechanism of self-alignment by the anisotropic photoexcitation within the plasma. For the same range of pressures, no radial symmetry of self-alignment of 2p states was observed in E-discharge (Fig. 42). The degree of polarization was the same for the central and the boundary regions of the plasma. Moreover, a strong polarization degree was observed under pressures lower than 0.1 Torr, for which no significant self-absorption on 1s - 2p transitions takes place, so self-absorption could not account for the self-alignment of 2p neon states at low pressures in E-discharge. Therefore even a simple qualitative experiment shows that the self-alignment of 2p neon states in the capacitive high frequency dis-

1 2

Figure 42. **Distribution of the self-alignment axes of the excited states of noble gases in a capacitive high frequency discharge under radiation self-absorption (1) and electron impact (2)**

charge is a complex effect, incorporating resonance radiation absorption and collisional excitation. Taking into consideration the complex nature of self-alignment in the capacitive high frequency discharge, it is possible to consider that the degree of polarization of the emission spectral lines depends on the quadrupole moment of the velocity distribution function $f_{2+}^{(2)}(\xi)$, the local anisotropy of the the resonance optical field $I_{2+}^{(2)}$ and the absorption coefficient of the spectral line connecting the levels studied with the low-lying states. It is expressed using formulae (1.35), (1.39) and (1.45) as:

$$\begin{aligned} P &= \frac{\omega(JJ_1)}{I_0}\frac{\Gamma_0}{\Gamma_2}\frac{8\pi}{5m^2}\int_{\varepsilon_t}^{\infty} d\varepsilon\, \varepsilon\sigma^{(2)}(\varepsilon) f_{2+}^{(2)}(\varepsilon) \\ &+ \frac{\omega(JJ_1)}{I_0}\frac{\Gamma_0}{\Gamma_2}\frac{\sqrt{6}}{5N_a}\sum_i \frac{B(\omega_i)c\hbar}{\omega_i}(-1)^{J+J_1}\left\{\begin{array}{ccc} 1 & 1 & 2 \\ J & J & J_i \end{array}\right\}\chi_i I_q^{(\kappa)}(J_iJ) \end{aligned}$$

It is easy to conclude that the study of the physical reasons for the polarization of spectral lines should involve a number of supplementary experiments: analysis of the self-alignment signal shape for different locations $P(H,x)$, measurement of the absorption coefficients for the spectral lines in the plasma χ_i, irradiation of the discharge in order to change the anisotropy of the optical field within the plasma $I_{2+}^{(2)}(J_iJ)$ and use of the reference electron impact excitation in the electron beam. Spherical discharge with a high degree of symmetry is most useful for such supplementary experiments because radial distribution of the alignment tensor $^{\vec{r}}\rho_0^{(2)}(r)$ may be expected for uniform distribution of the plasma emission. For the low absorption coefficient, the character of the anisotropy of the resonance optical field within the plasma is determined by the discharge shape. The multipole moments of the resonance light exictation tensor in this case according to (1.37) and (1.39) are given by the approximate formulae valid for the qualitative examination of the self-alignment distribution within different discharges:

$$\breve{F}_q^{(\kappa)} = \sum_i \tilde{\mathcal{F}}_\kappa(JJ_i) \int d\vec{n}\, C_q^{(\kappa)^*}(\vec{n}) L(\vec{n}) \bar{\varepsilon}_i \tag{4.1}$$

For the spherical discharge

$$L(\vec{n}) + L(-\vec{n}) = 2\sqrt{R^2 - r^2 \sin^2\theta},$$

where R is the radius, r is the direction from the center to the studied point and θ is the angle between the vector $\vec{n}$ and the radius-vector of this point /22/. The longitudinal component of the alignment tensor (1.37) will take the form:

$$\rho_0^{(2)}(r) = \sqrt{\frac{3}{2}} \frac{c\hbar}{\Gamma_2 N_a} \sum_i \frac{B(\omega_i)\bar{\varepsilon}_i}{\omega_i} (-1)^{J+J_1} \begin{Bmatrix} 1 & 1 & 2 \\ J & J & J_i \end{Bmatrix}$$

$$\times_i \int_0^{\pi/2} d\theta \sin\theta (3\cos^2\theta - 1)\sqrt{R^2 - r^2\sin^2\theta} \tag{4.2}$$

where $\bar{\varepsilon}_i$ is the local spectral line emission efficiency.

Such radial dependence of the alignment tensor gives monotonous increase of the polarization degree $P(x)$ from the zero value in the center of the discharge and different signs of the polarization degree for the regions of observation located along the horizontal and vertical radii (Fig. 43). The distribution of self-alignment for the electron impact process should be determined by the distribution of the high frequency field. It means that for the capacitive discharge the sign of the degree of polarization should be the same for all observed locations. Distribution of the degree of polarization across the discharge image was studied for 1s - 2p spectral array in noble gases Ne, Ar, Kr /65/. Strong pressure dependence on the distribution of self-alignment was observed. At a pressure higher than 4 Torr the self-alignment degree for all spectral lines was very low due to depolarizing collisions. At pressure below 1 Torr the distribution was radially symmetric. The variation of the experimental profiles and the signs are presented in Fig. 42. Dispersive contours registered in the diagonal points indicated the

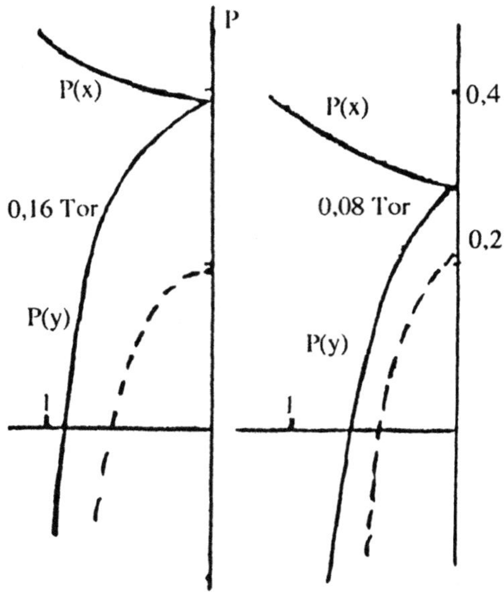

Figure 43. **Distribution of Stokes parameters P(x), P(y) for 811.5 nm argon line emitted by a spherical high frequency capacitive discharge for different pressures without (full lines) and with a mirror (dashed line)**

inclination of the alignment axes at a 45°angle with respect to the horizontal axis. Radial symmetry of distribution of the alignment axes within the boundary regions and the characteristic variation of the degree of polarization while scanning to the center of the discharge image (Fig. 43) showed that radiation self-absorption was the reason for the self-alignment of 2p states. Experimental estimates of $\chi_0 R$ parameters for those experimental conditions ($\chi_0 R = 0.5 - 4.5$) for different 1s - 2p spectral lines confirmed that radiation self-absorption mechanism is important in this case.

The nonzero polarization degree $P(x)$ in the center of the discharge, which is not accounted by formula (4.2), was due to the non-spherically symmetric distribution of the excited atoms in the discharge. This idea was confirmed by the observation of different intensity profiles along radial and vertical axes of the discharge. It means that atoms in the center of the discharge are irradiated non-isotropically, being the reason for the non-zero self-alignment effect on 2p states in the center of the discharge. By mounting a horizontal mirror to the lower part of the discharge, the optical anisotropy in the center was partially compensated and a decrease in the degree of polarization from the center was observed in all cases. Moreover, all polarization profiles shifted down (Fig. 43) to be closer to the theoretical distribution (4.2).

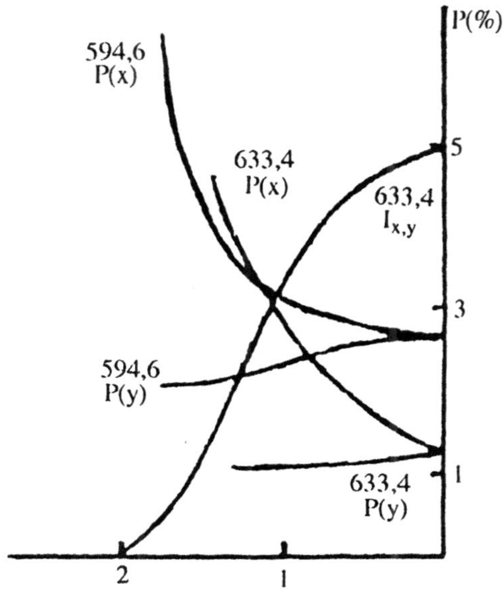

Figure 44. **Distribution of I and P Stokes parameters for 633.4 nm neon line (80 mTorr) and 549.6 nm argon line (17 mTorr) emitted by a high frequency spherical discharge**

4.1.2 ELECTRON IMPACT

A number of dramatic changes in the spectropolarimetric characteristics have been observed in the emission of high frequency discharge at pressures below 0.1 Torr for neon and 0.05 Torr for argon /65/. The intensity distribution of the lines on 1s-2p transition array of noble gases was approaching isotropic (Fig. 44), and absorption coefficients for line centers decreased sharply, but the degree of polarization increased in all parts of the discharge. Polarization profiles were also changed and approached the dependence presented in Fig. 44. Magnetic depolarization signals had the same sign for all discharge locations. These observations indicated that alignment axes became preferentially oriented along the high frequency field direction (Fig. 42). These results may be explained assuming that at low pressures the efficiency of photoexcitation of 2p states decreases and that self-alignment is mainly determined by electron impact excitation anisotropy. No changes have been observed in this case when a supplementary mirror was used. The self-alignment of the highly excited s and d states at low pressures, as estimates of the absorption coefficients show, can only be only due to the electron impact process. Polarization profiles for the highly excited states are principally different with respect to those for 2p states at higher pressures (Fig. 43). An important experimental proof for the impact self-alignment process was the comparison of the polarization effects in the central part of the plasma with the electron beam experiment, assuming that the symmetry of the electron impact excitation in the central part is caused by the oscillating electric field oriented perpendicular to the plates of the capacitor. The localisation of the magnetic spectropolarimetric signal shapes and signs in the plasma in the collision chamber of the electron beam was observed only for the low pressure range, confirming the electron impact self-alignment mechanism. For pressures higher than 1 Torr, no correlation was present.

4.2 Plasma structural features in a capacitive discharge

Polarization profiles of high frequency discharge even on the qualitative level, represent a very powerful means for the sensing of different characteristics of the plasma which reflect the structural features of the ionized gas as an integrated entity.

4.2.1 RADIATION POLARIZATION IN THE VICINITY OF THE ELECTRODE SHEATH

Electrode spectropolarimetric effects have been studied in the cylindric high frequency discharge. The special role of the electrode boundary layers of plasma was reflected in the axial intensity distribution. For pressures lower than 0.1 Torr a dark electrode space of several millimeters was observed. It broadened when the pressure decreased. For pressures lower than 15 mTorr (for argon), the distribution of intensity of spectral lines was monotonous, increasing to the center of the discharge. For higher pressures the maximum broadened and shifted to the center of the discharge while pressure decreased. Spectral line intensity increased in all points of the discharge with the growth of the high frequency voltage.

The sharp polarization maximum positioned close to the boundary of the electrode dark space was observed for all spectral lines in the axial distribution of linear polarization (Fig. 45) /66/. At low pressure, the electrode polarization maximum became more sharp and shifted to the center of the discharge. Out of the electrode polarization maximum, the smooth monotonous variation of the polarization profile was typical for all spectral lines. For all the noble gases there was no energy selectivity of self-alignment which was typical for the positive column of the gas discharge. Electrode polarization maxima in all cases increased with the applied high frequency field. Study of the distribution of the shape of magnetic depolarization signals across the discharge image allowed to represent the spacial distribution of self-alignment axes typical for the electron impact process, which

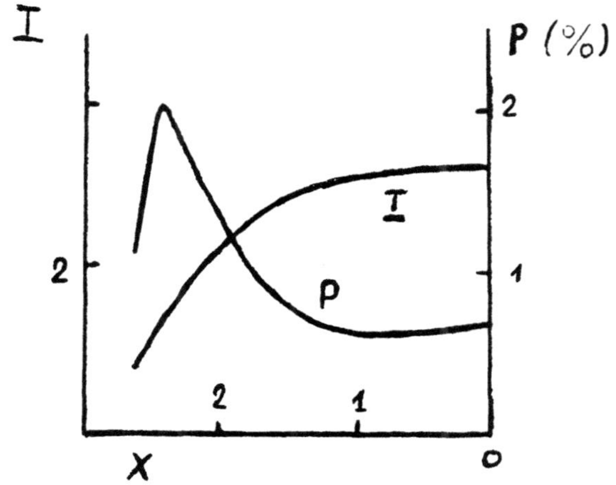

Figure 45. Polarization (P) and intensity (I) profiles for 603.2 nm argon line for a cylindric high frequency discharge

is given in Fig. 42.

4.2.2 ELECTRON MOTION ANISOTROPY NEAR THE ELECTRODE SHEATH

In order to explain spectropolarimetric observations of the boundary electrode regions of a capacitive high frequency discharge /67, 68/, a numerical experiment was implemented. As the mean free path of electrons was smaller than the characteristic length of the electrode sheath, a one-dimentional collisionless model of semi-infinite plasma with the infinite plane dielectric electrode was used. An electric field was directed perpendicular to the electrode along the OY axis, the electron concentration was taken equal to 10^{11} cm^{-3}, the electron temperature was $V_e = 4$ eV, and the Debye radius $d_e = \sqrt{V_e/4\pi e N_e} = 0.5$ mm. For these conditions, the electron and ion plasma frequencies of the noble gases were $3 \cdot 10^9$ and 10^2 Hz, so $\omega_i \ll \omega_\sim \ll \omega_e$. This means that the electron component was mobile enough to react to the temporal change of the high frequency electric field but the kinetic of ions was determined by the stationary electric field within the plasma.

The numerical analysis of electron kinetics in the boundary electrode region was performed as follows. A two-dimensional phase space (y, v_y) was divided into 128x128 cells and a certain value of the distribution function $f(y, v_y)$ was given to every cell. The Liouville - Vlasov equation:

$$\frac{df(v_y, y)}{dt} = 0$$

was solved numerically, implementing the differentiation along the electron trajectory in the phase space. The width of the cells in the phase space was made at every step of the programme, proportional to the velocity in the vertical direction and acceleration in the horizontal one: $a = eE(y)/m$. The local electric field was determined by solving the Poisson equation:

$$\frac{\partial E(y)}{\partial y} = 4\pi e(N_i - N_e)$$

where N_e and N_i are electron and ion concentrations.

Ion concentration was assumed to be constant and not dependent on time. Electron concentration was determined as the sum of all values of the velocity distribution function for all cells of the horizontal line, corresponding to a given y - coordinate. The initial electric field on the electrode was determined to be $E(0) = 4\pi\eta$ where η is the charge density on electrode. The quantity η was modulated by the external electrode current with account for the current of the plasma ions and electrons penetrating the electrode potential barrier. The phase space for different moments is presented in Fig. 46. The cells of the phase space are defined by the points which for $y > L_0$ (L_0 is the boundary electrode sheath width at the moment of the crossing of the electrode sheath boundary corresponded to velocities proportional to $v_T/2$ ($v_T = \sqrt{2eV_e/m}$ is the thermal electron velocity). The velocity distribution of the electrons incident on the boundary layer was approximated by a Maxwellian function

$$f(v_y) = \frac{N_e}{\sqrt{\pi} v_T} \exp\left(-v_y^2/v_T^2\right), \qquad (v_y < 0)$$

These points were moving within the phase space. The picture for different steps is given in Fig. 46.

The first stages of calculations were performed with no electric field applied. The initial electric charge on the dielectric wall was zero. Calculations showed that an electrode boundary sheath is forming with a well defined boundary, preventing electrons from entering the sheath. This result corresponds to the general concept of formation of the boundary potential profile. Due to the high mobility of electrons, the electron current to the wall initially exceeds the ion current and the electric field repulsive to electrons increases. The electron and ion currents balance out and the boundary potential barrier reflects the majority of the bulk plasma electrons.

The external electric field with the frequency $\omega_\sim \ll \omega_e$ applied to the electrode is located within the thin electrode sheath. As seen in the Fig. 46, a boundary layer is also formed between the quasineutral plasma and the positively charged electrode sheath where there are practicaly no electrons. The decrease in electron concentration from the mean bulk plasma value

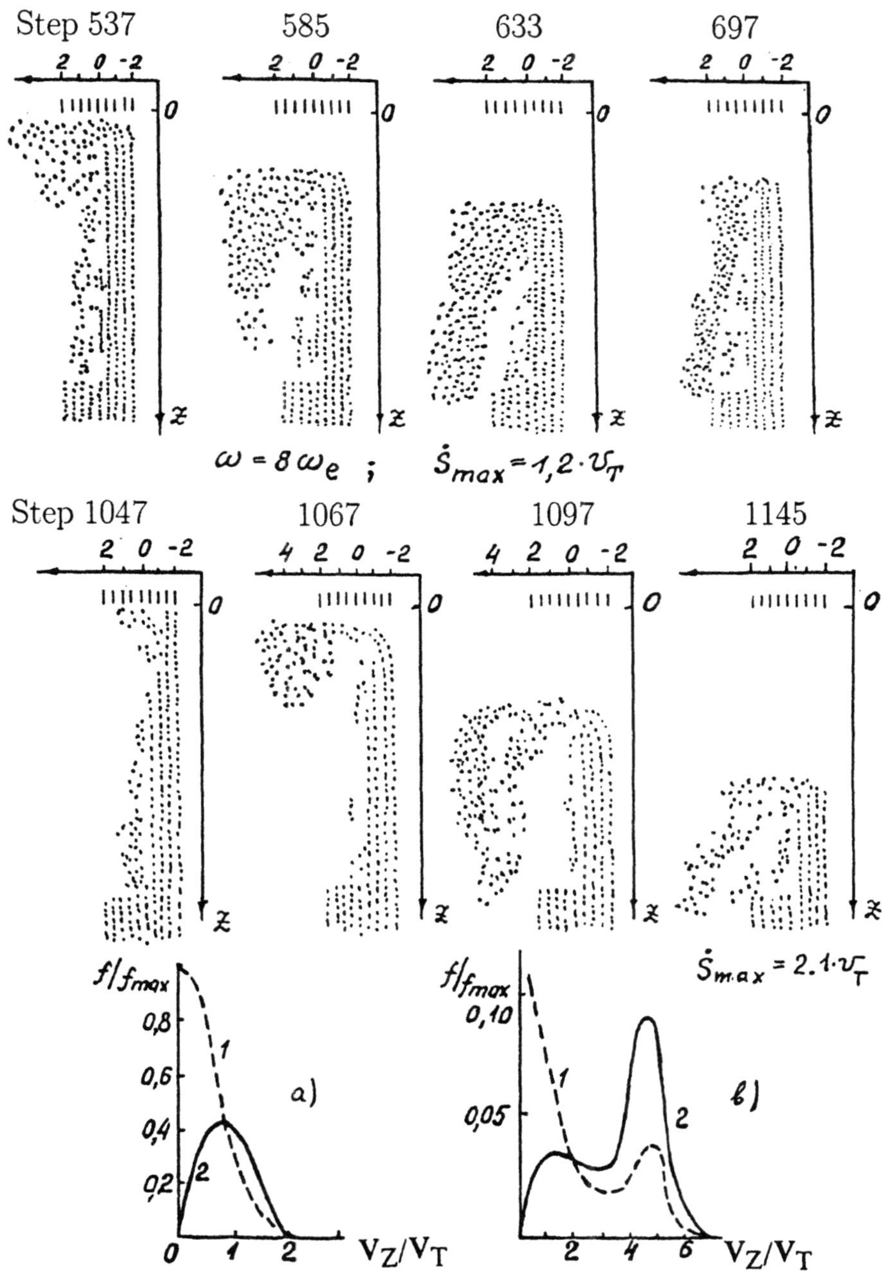

Figure 46. Electrode sheath in a capacitive high frequency discharge and formation of the velocity distribution function of reflected electrons. Phase diagrams for the electron component at different steps of the simulating experiment as a function of z and of the $\dot{L}/v_T$ parameter. Maxwellian distribution function (1) and flux (2) for incident (a) and reflected (b) electrons for $\dot{L} = 2.1\, v_T$, (v_T is the thermal velocity)

to zero takes place on a distance of the same order as Debye length (d_e) that is smaller than the electrode sheath width.

In order to estimate the electrode sheath width, let us examine the one-dimensional Poisson equation:

$$\frac{\partial^2 \varphi}{\partial y^2} = 4\pi e(N_i - N_e) \tag{4.3}$$

Assuming that $N_i = N_0 = \text{const}$, $\varphi(y) = 0$ out of the boundary sheath and approximating the distribution of electrons by the step function ($N_e = 0$ for $0 < y < L$, $N_e = N_0$ for $y > L$, where L is the instantaneous coordinate of the sheath boundary) we have:

$$\varphi(y) = 2\pi e N_0 (y - L)^2 \ (y < L)$$

The quantity L may be estimated using the condition $\varphi(0) = V$, (V is the instantaneous value of the electrode-plasma potential difference):

$$L = \sqrt{\frac{V}{2\pi e N_0}} = \sqrt{2\frac{V}{V_e}} d_e$$

So, for $V_{\sim} = 200$ V, the maximum width of electrode sheath $L_0 = L_{max} = 10 d_e$, which is in agreement with experimental data.

The numerical experiments showed that the boundary of the electrode sheath is like a rigid wall, reflecting the bulk electrons. Indeed, near the boundary of the electrode sheath, the plasma potential at a distance of a Debye length increases from zero to a value far exceeding the thermal electron energy V_e. This means that the bulk plasma electrons are reflected directly from the boundary, and not penetrating into the sheath. Therefore the investigation on the electron velocity distribution anisotropy may be carried out regarding the electrode sheath as a thin positively charged layer with an oscillating boundary which presents a potential profile with an effective length of the potential drop much lower than the sheath width.

The second important result of the numerical experiments was the formation of an ensemble of reflected electrons with a mean velocity exceeding the thermal velocity of the bulk electrons. In other words, This may be interprted as the formation of a pulsed quasi-electron beam with a mean energy exceeding the mean energy of the bulk plasma electrons. The distribution function averaged over the period of the field deviates from a Maxwellian distribution function and possesses the characteristic hump of fast beam electrons (Fig. 46).

Simple kinetic considerations may be applied to the analysis of the general character of the velocity distribution of electrons reflected from the

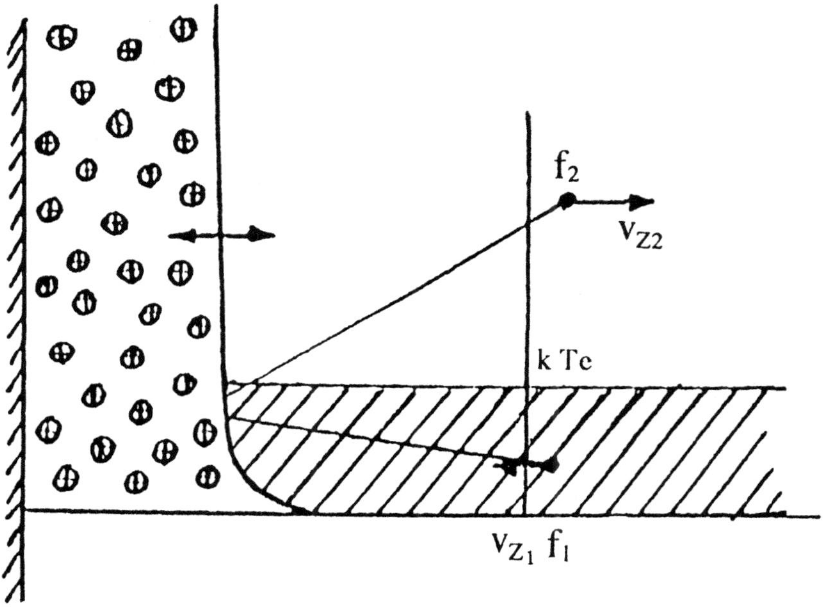

Figure 47. Electrode sheath model of a capacitive high frequency discharge

oscillating boundary. In a frame of reference connected to the electrode sheath boundary, the reflection is elastic and after transformation to the laboratory frame of reference the velocity of a reflected electron will be $v_{y2} = -v_{y1} + 2\dot{L}$, where $v_{y1,2}$ are the y-components of the electron velocity before and after the reflection, and L is the instantaneous sheath boundary velocity. In the vicinity of the boundary, the electron velocity distribution function is made of two parts: the velocities distribution of the incident ($v_{y1} < 0$) and reflected electrons ($v_{y2} > 0$) (Fig. 47). According to the Liouville theorem, in an electric field, the velocity distribution function does not change its value while moving along the electron trajectory in the phase space:

$$f_1(v_{y1}, t_1) = f_2(v_{y2}, t_2),$$

where $t_{1,2}$ are moments when an electron crosses the sheath boundary and flies out of it. Assuming the distribution of the incident electrons to be stationary, and averaging over the field period, we may write the following expression for the velocity distribution function of reflected electrons :

$$\overline{f_2}(v_{y2}) = \frac{1}{T} \int_0^T dt_2\, f_1\left[v_{y1}(v_{y2}, t_2)\right], \tag{4.4}$$

where at the right hand side the distribution function of incident electrons is represented as a complex function of v_2, t_2. In this manner the calculation of the velocity distribution of the reflected electrons is reduced to the determination of the trajectories in the reflection on the oscillating boundary layer.

For a saw-tooth law for $L(t)$, which applies to a number of experiments, it is possible to obtain an analytical representation of $f_2(v_{y2})$ (4.4). In the case $v_{y2} < u$ ($u = |\dot{L}|$), all incident electrons had the velocity

$$v_{y1} = -v_{y2} - 2u \quad \text{and} \quad f_2(v_{y2}) = f_1(-v_{y2} - 2u).$$

Depending on t_2, for the case $u < v_{y2} < 2$ two types of electron trajectories within the sheath are possible. Solving the simple kinetic problem for the elastic reflection of electrons on the sheath boundary, we have for the velocity distribution function the following expression:

$$f_2(v_{y2}) = \frac{1}{2}\left(1 + \frac{u}{v_{y2}}\right) f_1(-v_{y2} - 2u) + \frac{1}{2}\left(1 - \frac{u}{v_{y2}}\right) f_1(-4u + v_{y2}) \quad (4.5a)$$

If $2u < v_{y2} < 3v$, three possibilities exist and the distribution function is represented as:

$$\begin{aligned} f_2(v_{y2}) \;=\; & \tfrac{1}{2}\left(1 - \tfrac{u}{v_{y2}}\right) f_1(-v_{y2} - 2u) + \tfrac{1}{2}\left(\tfrac{3u}{v_{y2}} - 1\right) f_1(-4u + v_{y2}) \\ & + \left(1 - \tfrac{2u}{v_{y2}}\right) f_1(-v_{y2} + 2u) \end{aligned} \quad (4.5b)$$

for $v_{y2} > 3u$ we have:

$$f_2(v_{y2}) = \frac{1}{2}\left(1 + \frac{u}{v_{y2}}\right) f_1(-v_{y2} + 2u) + \frac{1}{2}\left(1 - \frac{u}{v_{y2}}\right) f_1(-v_{y2} + 2u) \quad (4.5c)$$

Fig. 48 shows the distribution function calculated using these expressions for a Maxwellian approximation of $f_1(v_{y1})$:

$$f_1(v_{y1}) = (N_e/\sqrt{\pi} v_T) \exp\left(-v_{y1}^2/v_T^2\right)$$

and different relations between u and v_T. For $u \geq v_T$, a characteristic maximum in the fast electron region takes place. It may be explained as a beam of electrons directed to the center of the discharge with the velocity $v_b = v_T + 2u$ associated with the saw-tooth $L(t)$ dependence. This mechanism of electron heating in the boundary electrode region is analogous to the Fermi process of fast electron generation in outer space /69/. Observation of the strong anisotropy of fast electron motions near the electrode sheath was carried out by the electric probe technique /70, 71/, and a beam of fast electrons was detected. For low pressures of noble gases the

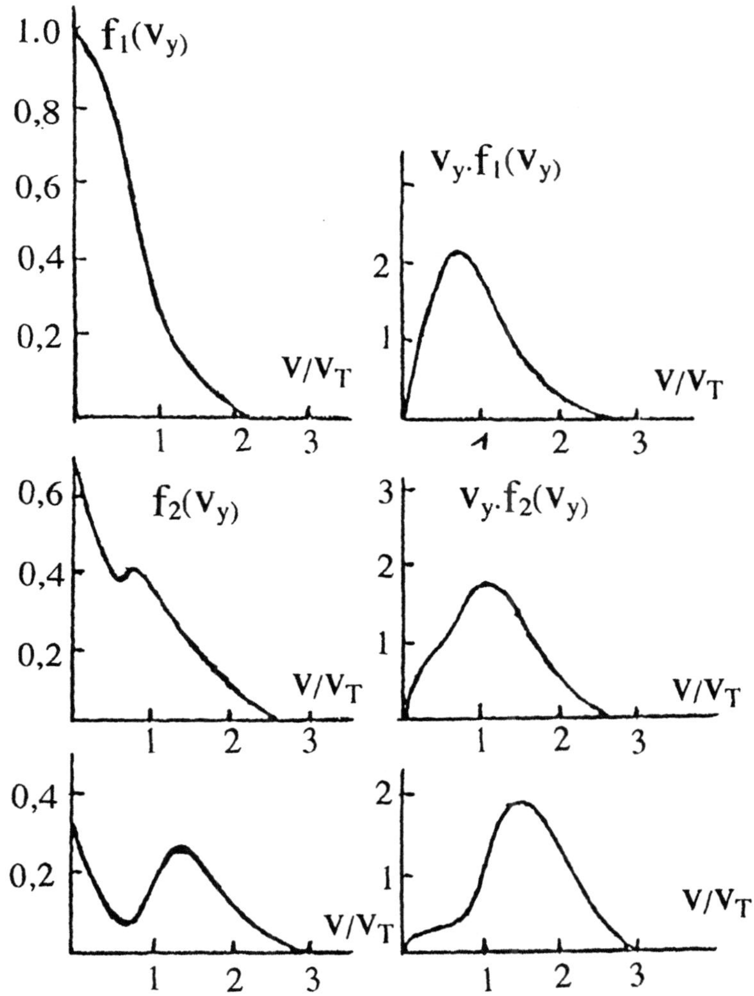

Figure 48. **Distribution function of v_y components, $f(v_y)$ and electron flux $v_y f(v_y)$ in the vicinity of the electrode sheath of a high frequency capacitive discharge calculated in Fermi model for different L and v_T parameters: 1 - Maxwellian distribution function and electron flux, 2 and 3 - distribution function and flux of reflected electrons for $\dot{L} = 0.3\, v_T$ and $\dot{L} = 0.5\, v_T$**

electron beams originating from the electrode sheath have been observed visually /72/. In a cylindrical discharge cell of 6 mm length, at frequencies 33 and 100 MHz, and at a pressure of 80 mTorr, the dark electrode space was rather wide and the distribution of the emission intensity presented bright cone regions on the background of the relatively weak diffuse

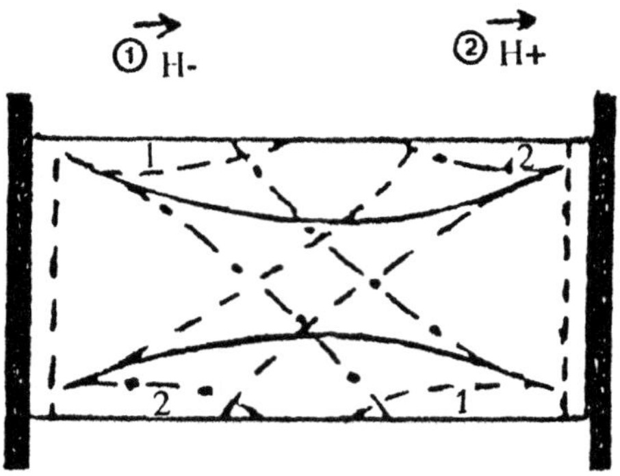

Figure 49. **Observable picture of the beam localization in a high frequency discharge without a magnetic field (full line) or with a magnetic field towards (dashed lines(1)) to or outwards (dashed-dotted lines (2)) from the observer**

emission of the other parts of the discharge (Fig. 49). The same structure of radiation was observed for the saw-tooth and the sinusoidal temporal dependence of the high frequency voltage applied. In the weak magnetic field, the side surface of those cones had been curved, shifting to different sides in the plane perpendicular to the magnetic field vector (Fig. 49). The magnetic field effect brought about the conclusion that these bright regions were formed by the beams of the fast electrons moving to the center of the discharge. The sharp side boundary of the beam region in the magnetic field indicated that the beam electrons were rather monokinetic. The distribution of velocities of the fast electrons was reflected in the distribution of the alignment tensor in the plasma. The study of the self-alignment signal shapes (Fig. 42) indicated that the local alignment axis were distributed parallel to the surfaces delimiting the electron beam. Electron beam observation confirmed the validity of the concept of formation of anisotropic fast electrons motions in a high frequency capacitive discharge plasma.

4.2.3 SPECTROPOLARIMETRIC EFFECTS IN OTHER PARTS OF THE PLASMA

The numerical study of the electron velocity distribution anisotropy in the boundary electrode regions enables one to analyse the spectropolarimetric data in connection to the structural features of the high frequency discharge, and to study the role of the different parts of the discharge in the processes responsible for the energy balance. The dark electrode space represents the positively charged sheath from which electrons are pushed out periodically by its oscillating boundary. A direct evidence for the ordering of the fast electron velocities near the boundary of the electrode sheath is the observation, for all spectral lines, of a sharp polarization maximum regardless the excitation potential of the line. The position of this maximum indicates the coordinate of the section of the discharge in which the electron beam is formed. The principal role of the electrode sheaths in the structural features of the discharge and the energy balance mechanisms were confirmed by the polarimetric studies of a discharge with a short interelectrode distance /67/. By shifting the internal electrode of the discharge used in the experiments, it was possible to change the distance between electrodes keeping the same discharge parameters. When electrodes approached each other, the bright region was transformed into a bright layer between two dark electrode sheaths of constant width, not depending on the interelectrode distance. Electrode polarization maxima approached each other, and for an interelectrode distance of 2 cm overlapped and were reduced to one maximum (Fig. 50). In order to analyse the electron motion anisotropy in the plasma at a small interelectrode distance, a one dimensional collisionless model was used. Spectropolarimetric data indicated that two interacting electrode sheaths must be considered in this case. Therefore, for the description of this discharge, the current of the second electrode was modulated with a phase opposite with respect to the current of the first one. The electric field of the second electrode was also within the boundary region of the first one. At the initial moment, the plasma between electrodes was assumed to have a Maxwellian velocity distribution function of electrons. The picture of the temporal variation of the phase space as a result of the numerical calculation is presented in Fig. 51. Oscillating electrode sheaths form with a quasineutral plasma in between. Oscillations of electrons take place at the frequency of the high frequency field. The mean distribution function of the y-velocity components remains Maxwellian but is shifted to the velocity of the boundaries of the electrode sheaths. The temporal dependence of the surface charge density on one electrode $\eta(t)$ and the potential difference between electrodes $V(t)$ are given in Fig. 51. For some initial time interval the $V(t)$ dependence shows oscillations at the Langmuir frequency. In order to explain this effect, let us calculate the impedance of the discharge, assuming a stepwise distribution of the electron concentration: $N_e = 0$ within the sheath and $N_e = N_0$ for the central part. The potential difference between electrodes, $V(t)$, consists of the electrode sheath potential drops, $V_{1,2}$, and the plasma potential distribution V_p :

$$V = V_1 + V_2 + V_p = 2\pi e N_0 (L_1^2 - L_2^2) - E_p L_p \tag{4.6}$$

where E_p is the electric field strength in the center of the discharge:

$$E_p = 4\pi(\eta_1 + eN_0 L_1)$$

here η_1 is the charge density on the first electrode.

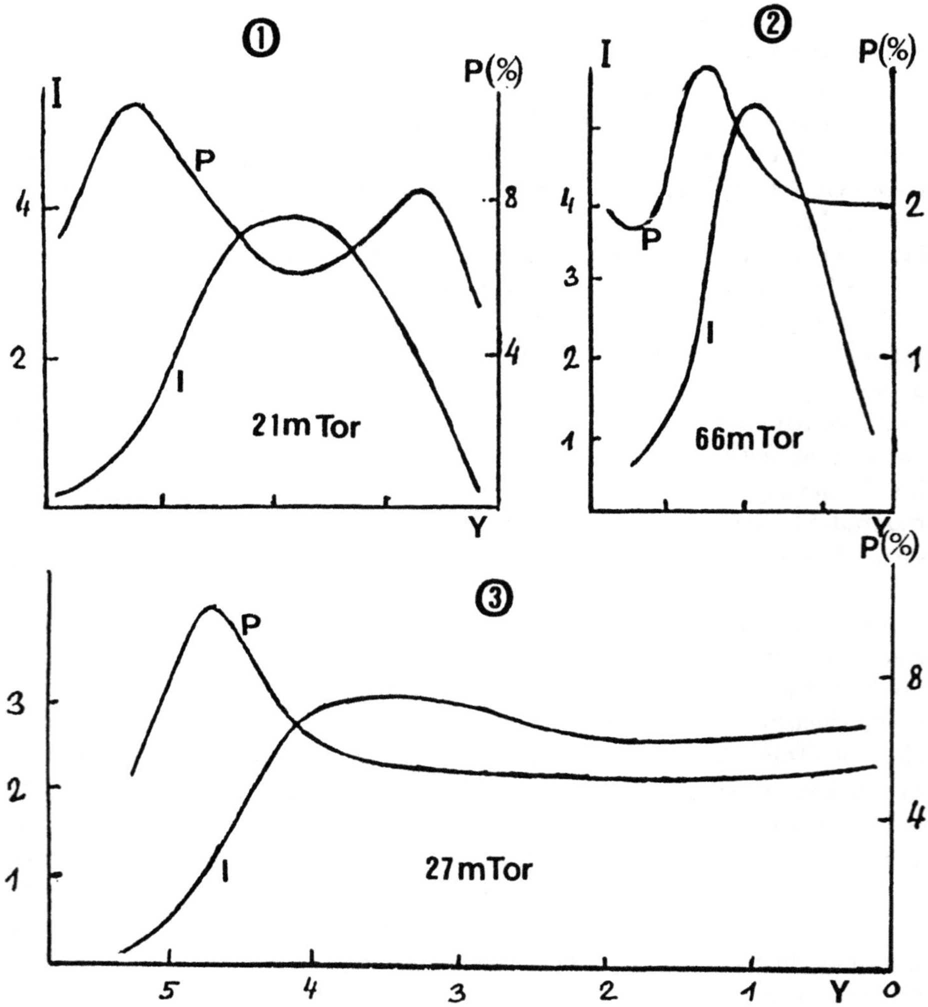

Figure 50. Polarization profiles of the 492.2 nm helium line as a function of the longitudinal coordinate for a cylindric high frequency discharge at different interelectrode distances: (1) - 3 cm, (2) - 2 cm, (3) - 12 cm

Assuming that $L_1 + L_2 = \text{const} = 2\bar{L}$ ($\bar{L}$ is the mean width of the electrode sheath) we have:

$$V = 8\pi N_0 e\bar{L}(l_1 - \bar{L}) - 4\pi(\eta_1 + eN_0 L_1)L_p.$$

If in the central region of the plasma the electron gas may be regarded

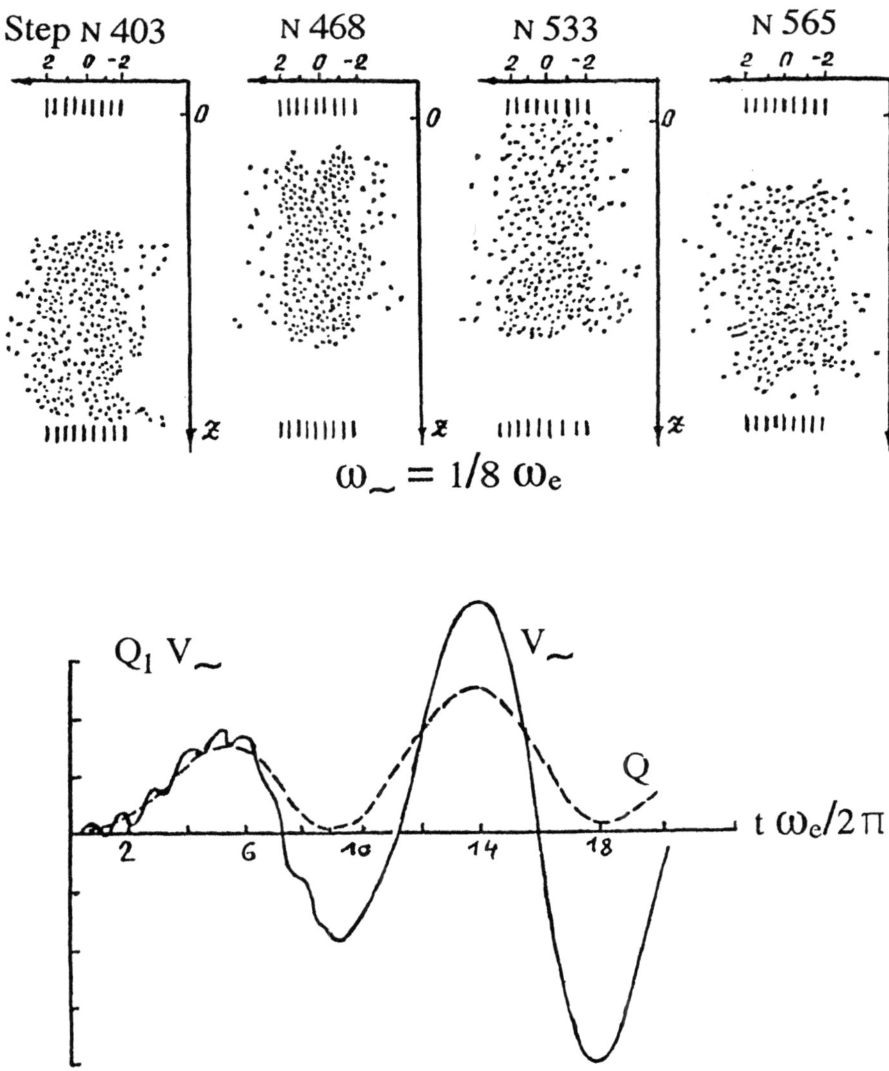

Figure 51. **Phase diagrams for the electron component at different steps of the numerical experiment, temporal dependence of the surface density charge (Q) and of the potential difference between electrodes ($V\sim$) for a high frequency capacitive discharge with a small interelectrode distance**

as oscillating in the high frequency electric field, we may write for the collision-free case the following equation:

$$\frac{\partial^2 L_1}{\partial t^2} = eE_p/m = 4\pi(e\eta_1 + N_0 e^2 L_1)/m$$

or, using the known temporal dependence of the discharge parameters:

$$L_1 = \bar{L} + L_\sim \exp(i\omega_\sim t) \qquad \eta_1 = \bar{\eta} + \eta_\sim exp(i\omega_\sim t)$$

we have

$$\begin{aligned} L_\sim &= 4\pi\eta_\sim(e/m)(\omega_\sim^2 - \omega_e^2) \\ \bar{L} &= -\bar{\eta}/N_0. \end{aligned}$$

The expression of the impedance of the high frequency discharge $z = V/\dot{\eta}$ then takes the form:

$$z = z_L + z_p = \frac{4\pi}{i\omega_\sim}\frac{2\omega_e^2\bar{L} - \omega_\sim^2 L_p}{\omega_\sim^2 - \omega_e^2}$$

where z_L is the total impedance of the electrode sheaths and z_p is the impedance of the central region of the plasma.

This expression indicates two types of resonances in high frequency discharges with a small interelectrode distance at low pressures. The first takes place at $\omega_\sim = 2\omega_e\bar{L}/L_p$ and corresponds to $z = 0$. The second resonance occurs at $\omega_\sim = \omega_e(z = \infty)$ and is the reason for the interelectrode voltage oscillations at Langmuir frequency (Fig. 51). When the electrode current is turned off, or the circuit is broken, the bulk electrons remain oscillating for some time, which leads to oscillations of the electrode voltage.

Taking into account the electron-atom collisions, the expression for the impedance takes the form:

$$z = \frac{4\pi}{i\omega_\sim}\frac{2\bar{L}\omega_e^2 - (\omega_\sim^2 - i\omega_\sim v)L_1}{\omega_e^2 - \omega_\sim^2}.$$

These numerical calculations of the electron motion anisotropy in a high frequency discharge for small interelectrode distance enable one to use the spectropolarimetric effects in the emission spectrum for the study of the parameters of this discharge. The quadrupole moment of the velocity distribution function in this case is connected to the velocity of the boundary of the electrode sheath:

$$f^{(2)}(v) = \frac{5N_0}{4\pi^{3/2}v_T^3}\exp\left[\frac{v^2 - \dot{L}_1^2}{v_T^2}\right]\int_{-1}^{1} d\alpha\,(3\alpha^2 - 1)\exp(-2\alpha v\dot{L}_1/v_T^2). \quad (4.7)$$

The parameter $\dot{L}_1$, which may be measured on the basis of the spectropolarimetric data, is used to determine the energy transport into the discharge plasma. This information is of principal importance for the construction of the high frequency discharge spectral lamps used in a wide range of physical and technical applications.

Spectropolarimetric measurements allow one to study the momentum relaxation of the beam electrons while they pass to the center of the discharge from the electrode sheaths /66/. Observations in the cylindric cell show that for different helium, neon and argon spectral lines, for a pressure of 2-80 mTorr ($\omega_{\sim} = 30, 100 MHz, V_{\sim} = 30 - 250V$), the decrease of the polarization degree moving to the discharge center occurs at the distance of 1 - 2 mm. Taking into consideration that the mean free path of electrons at these pressures lies in the range 3 - 100 mm, it is possible to conclude that the characteristic scale of the polarization decrease does not depend on the electron mean free path in the plasma. It means that the isotropization of the velocities of the beam electrons is determined not by collisions with neutral atoms but is the result of the collective processes of beam-plasma instabilities. Comparison with intensity profiles under the same conditions show that the characteristic length of the energy relaxation of the beam plasma electrons is much higher than that of the momentum relaxation.

Monotonous variation of the polarization degree is observed while moving to the center of the discharge cell and a constant polarization degree is detected on the central parts of the plasma at long interelectrode distances /67/. The increase of the interelectrode distance led to the corresponding increase of the central part of the plasma with a constant degree of polarization (Fig. 50). The general concept of electron velocity anisotropy in this region of the discharge appeared as a result of the observation of the polarization degree in the vicinity of the side dielectric walls of the discharge /63/. The monotonous variation of the degree of polarization moving to the discharge center was detected for 2p neon ($\varepsilon_t = 18.3 - 19.0$ eV) and argon ($\varepsilon_t = 13.0 - 13.4$ eV) states. Polarization profiles for the lines on transitions from the highly excited neon ($\varepsilon_t > 20.7$ eV) and argon ($\varepsilon_t > 15.0$ eV) states showed a strong wall polarization maxima with a degree of polarization 2 - 3 times higher with respect to the value in the discharge center (Fig. 52).

Compared to the electrode polarization maxima which take place for all spectral transitions, the energy selectivity of the wall polarization maxima for the highly excited states is a principally different effect . Analogy with the positive column of the d.c. discharge allows one to consider that the wall polarization maxima in the high frequency discharge are determined by the formation of stationary loss cones in the velocity destribution function of fast electrons near the side dielectric walls. The transversal polarization

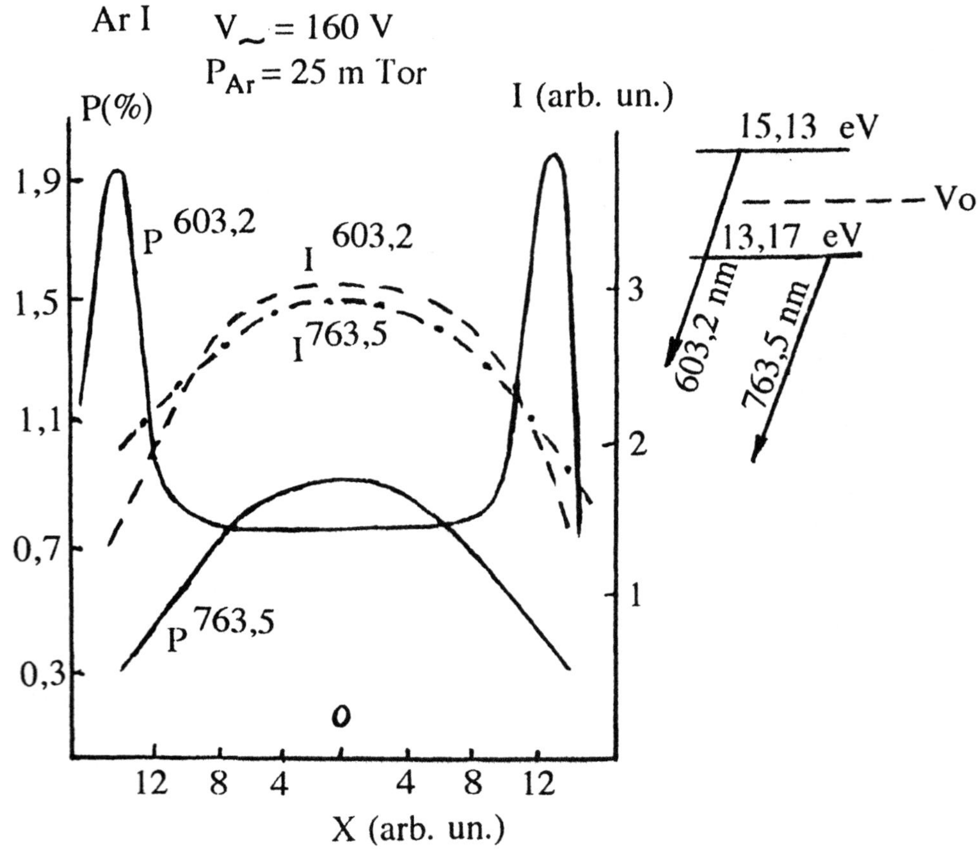

Figure 52. Upper level energy selectivity of the wall polarization maximum for a cylindric capacitive high frequency discharge for argon lines with different excitation potentials

profile for the highly excited states is connected to the variation of the characteristics of the loss cones and basically to the transversal stationary potential profile in the plasma. The self-alignment of the deep low energy states in the central part of the plasma is determined by the oscillation of electrons in the alternative electric field penetrating through the electrode sheaths.

4.3 Electric and energetic characteristics of a capacitive discharge

Analysing the spectropolarimetric features of the optical emission of the capacitive high frequency discharge for the range of frequencies $\omega_i < \omega_\sim < \omega_e$ it is safe to say that energy transport into a low temperature plasma is performed effectively within the electrode boundary region. This process is connected to the formation of beams of fast electrons propagating toward the centre of the plasma and forming sharp polarization maxima near the electrode sheath boundary. The energy balance in the central regions of the plasma is maintained by the alternative electric field penetrating through the electrode sheaths.

4.3.1 ENERGY TRANSPORT THROUGH THE ELECTRODE SHEATH

For the determination of the energy delivery into the plasma, the velocity distribution function of electrons, which is the energy-conducting component in the boundary sheath region, was represented by two components: incident and reflected electrons /73/. The Liouville-Vlasov approach has been utilized for deriving the velocity distribution function of reflected electrons near the nonstationary boundary, approximating the distribution function of the incident electrons with the help of known kinetic models of the bulk plasma electrons. Exerimental verification of this approach for such problems has been carried out on the first stage, making use of the experimental spectropolarimetric data in the cylindric argon discharge and of the simple model of a rigid oscillating wall for the saw-tooth temporal variation of its boundary.

Taking into consideration the low degree of self-alignment ($\rho_q^{(2)} \ll \rho^{(0)}$) for the second Stokes parameter of the emission line, we may write the following expression:

$$P = \frac{1}{5}\sqrt{\frac{3}{2}}w(JJ_1)\frac{\Gamma_0}{\Gamma_2}\int_{v_t}^{\infty} dv\, v^3\sigma^{(2)}(v)f^{(2)}(v) / \int_{v_t}^{\infty} dv\, v^3\sigma^{(0)}(v)f^{(0)}(v) \tag{4.8}$$

This integral equation for the known cross-section of electron-atom collisions and the zero-order multipole moment of the velocity distribution function, $f_0^{(0)}(v)$, enables one to obtain information about the quadrupole moment $f_0^{(2)}(v)$ which is proportional to the momentum flow

At the first stage the expressions for $f^{(0)}(v)$ and $f^{(2)}(v)$ have been obtained for the very simple model of elastic reflection of the falling Maxwellian bulk plasma electrons by the sheath boundary /66/ (see section 4.2). For the subensemble of the fast electrons responsible for the direct collisional excitation of the highly excited atomic states with velocites $v > 2u$ (u is the mean velocity of the electrode sheath boundary), the multipole moments (4.5) take the form /66, 74/:

$$f^{(0)}(v',\beta) = K_f\left[\exp\left(-v'^2\right) + \int_{2\beta/v'}^{1}\left(1 - \frac{2\beta}{v'\alpha}\right)\exp\left(4\beta v'\alpha - v'^2 - 4\beta^2\right)d\alpha\right]$$

$$f^{(2)}(v',\beta) = K_f \int_{2\beta/v'}^{1} (3\alpha^2 - 1)\left(1 - \frac{2\beta}{v'\alpha}\right) \exp\left(4\beta v'\alpha - v'^2 - 4\beta^2\right) d\alpha \tag{4.9}$$

where $\alpha = v_y/v$, $v' = v/v_T$, $\beta = u/v_T$ and K_f is a constant.

The measurement of the boundary polarization maxima has been carried out in a cylindric capacitive high frequency discharge of 4 cm diameter and 6 cm length, with outer plane electrodes, for argon lines 801.4, 763.5, 706.7 nm at a pressure of 27 mTorr, a capacitor voltage of 200 V and at a frequency of 100 MHz. For the same discharge conditions, the ratio of the relaxation constants Γ_2/Γ_0 was also measured by the Hanle effect technique (see Table 4.1).

TABLE 4.1. Measurements of the electrode polarization maximum P in a capacitive high frequency discharge and relaxation constants ratio for some ArI lines, and calculated β and u parameters for different thermal velocities v_T

λ, nm	Experiment		$v_T/10^8$.					
			1,0		1,2		1,4	
	Γ_2/Γ_0	P	β	u	β	u	β	u
801,4	1,029	0,0092 (9)	1,445 (6)	1,445 (6)	1,056 (8)	1,268 (9)	0,840 (10)	1,176 (14)
763,5	1,032	0,0094 (19)	1,417 (12)	1,417 (12)	1,037 (13)	1,244 (15)	0,831 (14)	1,164 (20)
706,7	1,032	0,0095 (19)	1,429 (13)	1,429 (13)	1,029 (17)	1,234 (20)	0,802 (20)	1,124 (29)

Using the electron impact alignment cross-sections calculated with the help of the differential cross-sections, electrode sheat velocity dependencies of the polarization, $P(\beta)$, have been obtained. The determination of the β parameter was carried out as a result of the comparison of the calculated dependencies $P(\beta)$ (Fig. 53) with the observed degree of polarization. Table 4.1 gives the results of the determination of the β parameter for every spectral line and for three values of v_T. Within the error bars, the coincidence between the values of the β parameter derived for different spectral lines confirms the reliability of this technique. The mean value of the sheath boundary velocity $u = v_T$ is given in Table 4.1 as well. For different v_T parameters, the difference in u is less than the uncertainty on v_T, which

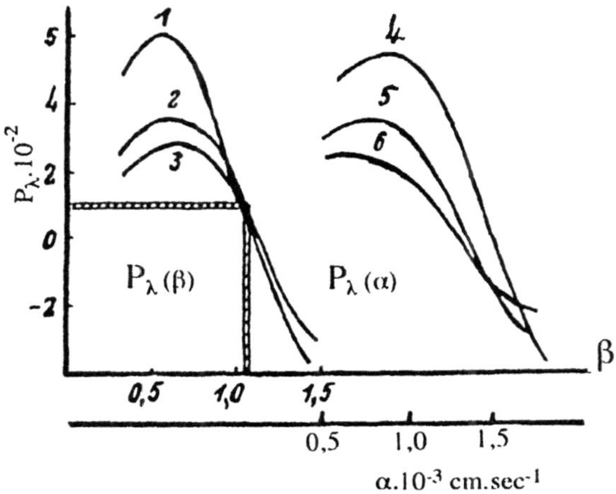

Figure 53. **Calculated reduced velocity dependence of the polarization in some argon line 1, 2 and 3 defined in Table 4.1**

was estimated according to /70/ as 15%. The reason for this is the low sensitivity of the function $P(u)$ to the parameter v_T (Fig. 53). By averaging all data for different λ and v_T, the velocity of the sheath boundary was determined to be $(1.3 \pm 0.2)10^8$ cm sec^{-1}. The correctness of this result was confirmed by comparison of the beam velocity $v_b = v_T/2 + 2u = 3.2(5) \cdot 10^8$ cm sec^{-1}, estimated from the sheath boundary velocity /66/ with the direct measurements of v_b by the magnetic field deflection technique /66, 72/. The difference between these values did not exceed the error bars of v_b from direct measurement. This is the second principal result confirming the possibility to utilize the spectropolarimetric sensing to study the energy transport into the plasma.

Within the approximation of an uniform ion distribution, the potential profile $\varphi(y,t)$ and the electric field strength at the electrode sheath are expressed by:

$$\begin{aligned} \varphi(y,t) &= 2\pi e N_0[y - L(t)]^2 \quad y < L(t) \\ E(y,t) &= -\partial\varphi/\partial y = 4\pi e N_0[L(t) - y] \\ \varphi(y,t) &= E(y,t) = 0, \quad y \geq L(t) \end{aligned} \tag{4.10}$$

where $L(t)$ is the instantaneous coordinate of the sheath boundary, N_0 is the ion concentration. The zero-point of the potential is the quasineutral bulk plasma.

The instantaneous electrode potential is given by $V(t) = \varphi(0,t) = 2\pi e N_0 L^2(t)$. If the impedance of the discharge considerably exceeds the exit resistor of the high frequency oscillator, we have, neglecting the small electrode-plasma potential drop :

$$V(t) = V_\sim(1 + \sin\omega_\sim t)$$

and

$$L(t) = \sqrt{V_\sim(1 + \sin\omega_\sim t)/2\pi e N_0} = L_0\sqrt{(1 + \sin\omega_\sim t)/2},$$

where $L_0 = \sqrt{V_\sim/\pi e N_0}$ is the boundary coordinate of the positively charged electrode sheath.

This regime is known as the potential regime. Otherwise, if the exit resistor is much higher than the discharge impedance, the current regime is implemented with a sinusoidal temporal variation of the discharge current and of the electrode charge. The width of the electrode sheath for the current regime is determined by the condition of compensating the volume positive charge by the electrode charge:

$$L(t) = L_0(1 + \sin\omega_\sim t)/2.$$

Calculations of the electron trajectories within the electrode sheath $y(t)$ for different v_{y2}, t_2 and $v_{y1}(v_{y2}, t_2)$ dependences have been carried out with the help of the equation:

$$m\frac{d^2y}{dt^2} = -eE(y,t),$$

in which the time and space of dependence $E(y,t)$ was obtained for different discharge regimes. Within the approximation of a soft high-frequency oscillating boundary, an electron may be inside the electrode sheath for the time interval exceeding the field period, interacting with the boundary several times. The phases of the free-flight and the accelerated motions are changing until the moment t_2 (Fig. 54). For the current regime, there exists an analytical solution to the equation of the motion:

$$y(t) = D_1 \sin(\omega_e t + \Phi) + D_2 \sin\omega_\sim t \tag{4.11}$$

where

$$D_2 = L_0\omega_e/2(\omega_e^2 - \omega_\sim^2),\ \omega_e = (4\pi e N_0/m)^{1/2} = 2V_\sim^{1/2}/L_0 m^{1/2}$$

and D_1 and Φ are constants determined by the initial conditions at the moment t' of the interaction of an electron with the sheath boundary :

$$D_1 = \left[(v_{y1} - D_2\omega_\sim \cos\omega_\sim t')^2/\omega_e^2 + (L(t') - D_2 \sin\omega_\sim t')^2\right]^{1/2}$$

$$\Phi = \arctan\left[\omega_e(L(t') - D_2\sin\omega_\sim t')/(v_{y1} - D_2\omega_\sim\cos\omega_\sim t')\right] - \omega_\sim t'$$

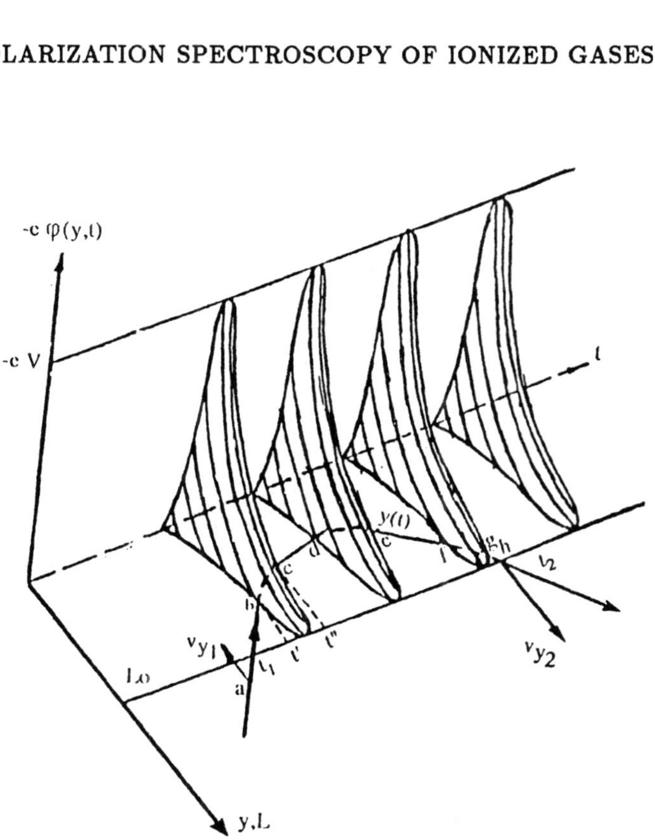

Figure 54. **Electron motion in the oscillating parabolic potential of the electrode sheath: free motion out of the positive space charge region and acceleration within this zone**

Numerical solution of the equation of motion may be used only for the potential regime. The calculated distribution function $\overline{f_2(\varepsilon)}$ is presented in Figs. 55 and 56. The maximum of the distribution function of the reflected electrons corresponds to electron beams propagating to the center of the discharge. Calculations of the degree of polarization have been implemented for 492.2 nm HeI line for a Maxwellian velocity distribution of incident electrons:

$$f_1(v_{y1}) = \int_{-\infty}^{\infty} dv_x \int_{-\infty} \lambda dv_z \, f_1(v_x, v_y, v_z) = N_e \pi^{-1/2} v_T^{-1} \exp\left(-\frac{v_{y1}^2}{v_T^2}\right).$$

At the boundary of the electrode sheath, the multipole moments of the velocity distribution function are expressed by :

$$f^{(0}(v) = \frac{N_e}{2\pi^{3/2} v_T^3} \left\{ e^{-\frac{v^2}{v_T^2}} + \frac{1}{T}\int_0^T dt_2 \int_0^1 d\alpha \exp\left[-\frac{v^2(1-\alpha^2) + v_{y1}(\alpha v, t_2)}{v_T^2}\right]\right\}$$

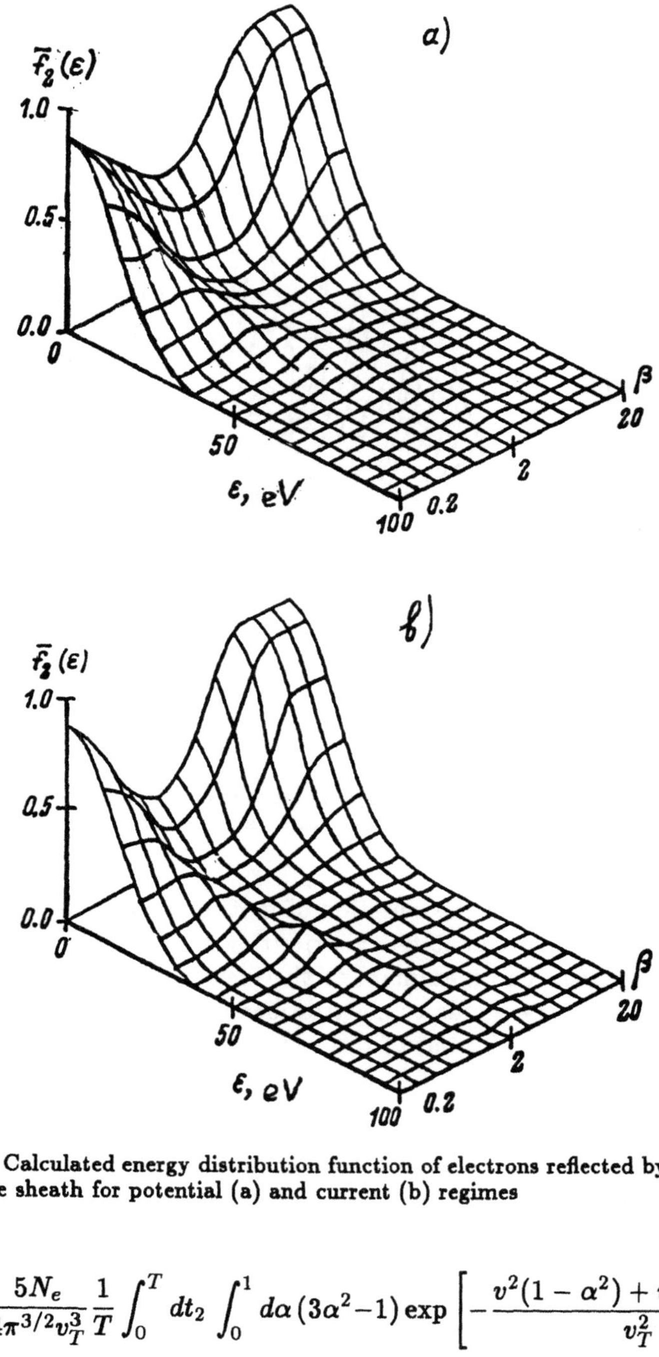

Figure 55. **Calculated energy distribution function of electrons reflected by the oscillating electrode sheath for potential (a) and current (b) regimes**

$$f_0^{(2}(v) = \frac{5N_e}{4\pi^{3/2}v_T^3}\frac{1}{T}\int_0^T dt_2 \int_0^1 d\alpha\,(3\alpha^2-1)\exp\left[-\frac{v^2(1-\alpha^2)+v_{y1}(\alpha v, t_2)}{v_T^2}\right], \tag{4.12}$$

where $\alpha = \cos\theta$ and θ is the angle $(\widehat{\vec{v}, OY})$.

The calculated degree of polarization at the electrode sheath boundary,

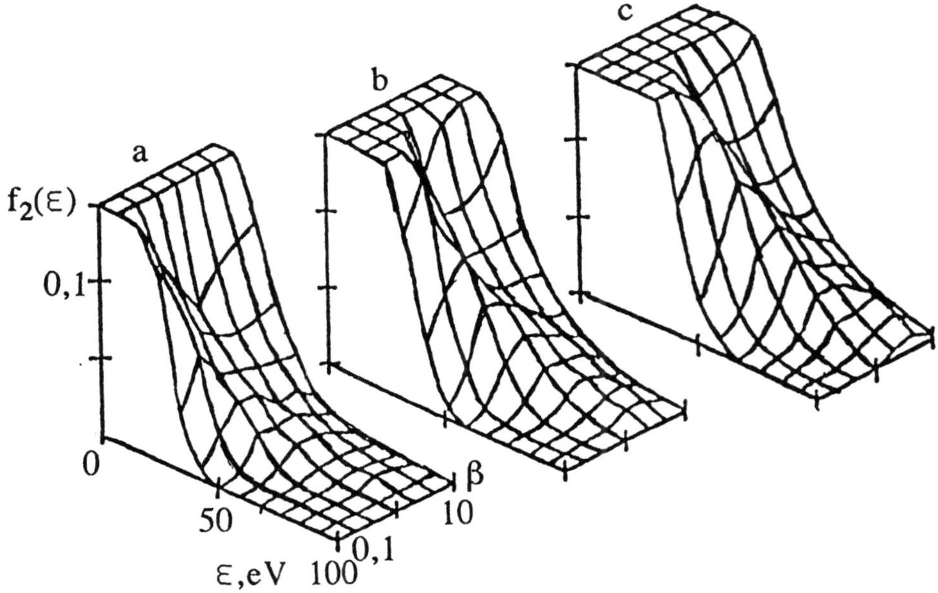

Figure 56. Calculated energy distribution function of the electrons reflected by the electrode sheath for the potential regime and for different electron temperatures: (a - 2 eV), (b - 4 eV), (c - 6 eV)

for the potential regime present in the discharge, is presented in Fig. 57 as a function of T_e and β parameters. Alignment and excitation cross-sections have been calculated using the atomic data from other independent works. The power transported into the plasma through the electrode sheath may be expressed as a result of the summation of the energy delivered to an electron over the total flux of electrons incident to the boundary sheath from the bulk quasineutral plasma and averaged over the period of h.f. field:

$$W = \frac{mS_e}{2T} \int_0^\infty dv_{y1} \int_1^T dt_1 \, [v_{y2}^2(v_{y1}, t_2) - v_{y1}^2] v_{y1} f_1(v_{y1}) \qquad (4.13)$$

The numerical calculation of the power delivered into the discharge plasma is given in the Fig. 57. The power is maximal for $u \sim v_T$ when the Fermi acceleration mechanism of the plasma bulk electrons takes place. The power increases with the electron temperature, reflecting the increase of the incident electron flow.

Spectropolarimetric determination of the power transported into the plasma was carried out for a capacitive discharge ($\omega_\sim$ = 100 MHz, $V_\sim$ =

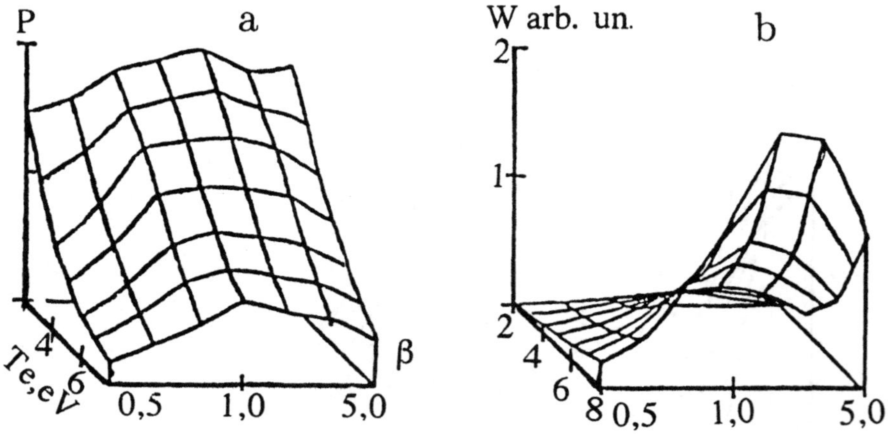

Figure 57. **Calculated dependence of the degree of polarization for the electrode polarization maximum (a) and of the power transported through the electrode sheath into the high frequency discharge plasma (b) as a functions of the discharge parameters β and T_e**

55 V, p_{He} = 18 mTorr). Supplementary measurements of the width of the electrode sheath have been implemented (L_0 = 7 mm), from which the mean electron concentration was evaluated ($n_e = 2.5 \cdot 10^8$ cm^{-3}). Using the spectropolarimetric data for the electrode boundary region, the electron temperature (T_e = 6(1) eV) has been determined by comparison with the calculated polarization degree. Finally using the expression (4.13), the power transported into the plasma was determined.

4.3.2 AMPLITUDE OF THE ALTERNATIVE ELECTRIC FIELD IN THE CENTRAL PART

The anisotropy of electron motions in the central part of the discharge is determined by the oscillative motions of electrons in the alternative high frequency electric field. This anisotropy is reflected in the polarization characteristics of spectral lines which are observed in the central part of the plasma. By analysing the kinetic equation, within the approximation of an uniform axially symmetric plasma in an alternative electric field, the multipole moments of the velocity distribution function of the plasma electrons may be obtained The nonstationary kinetic equation for the distribution function of electrons in an alternative electric field $E_y(t) == E_\sim \cos(\omega_\sim t)$

takes the form:

$$\frac{\partial f}{\partial t} + v_r \frac{\partial f}{\partial r} + eE_y(t) v_y \frac{\partial f}{\partial \varepsilon} + \frac{eE_y(t)}{\sqrt{2m\varepsilon}}(1-\alpha^2)\frac{\partial f}{\partial \alpha} = \Xi(f) \tag{4.14}$$

On the discharge axis, $\partial f/\partial r = 0$, $\alpha = \cos(\widehat{\vec{v}, OY})$ and ε is the kinetic energy of electrons. In this case equation (4.8) takes the form:

$$\frac{\partial f}{\partial t} + \frac{\partial f}{\partial v} a\alpha + \frac{\partial f}{\partial \alpha}\frac{a}{v}(1-\alpha^2) = \Xi(f)$$

where the variable $v = \sqrt{2m/\varepsilon}$ is used and $a = eE/m = a_\sim \cos(\omega_\sim t)$.

The kinetic equation incorporates the dependence on the direction of the velocity vector of electrons, so it is possible to study the angular anisotropy of the velocity distribution function.

Accounting for the axial symmetry of the discharge, the multipole representation of the velocity distribution function is expressed as:

$$f(v,\alpha) = \sum_{\kappa=0}^{\infty} f^{(\kappa)}(v) P_\kappa(\alpha)$$

where $f^{(\kappa)}(v) \equiv f_0^{(\kappa)}(v)$ and $P_\kappa(\alpha)$ are Legendre polynomials. The equation (4.8) will be transformed into:

$$\sum_{\kappa=0}^{\infty} \left\{ \frac{\partial}{\partial t} + a\alpha\frac{\partial}{\partial v} + a\frac{(1-\alpha^2)}{v}\frac{\partial}{\partial \alpha} \right\} f^{(\kappa)}(v).P_x(\alpha) = \sum_{\kappa=0}^{\infty} \Xi^{(\kappa)}(f) P_\kappa(\alpha), \tag{4.15}$$

where the multipole moments of the collision integral are represented by:

$$\Xi^{(\kappa)}(f) = \frac{2\kappa+1}{4\pi}\int d\vec{n}\, C_0^{(\kappa)*}(\vec{n})\Xi(f). \tag{4.16}$$

The collision integral for a low temperature plasma is described by expression (3.20). The Focker-Plank term, for atomic particles at low temperature of with respect to electrons, may be expressed as /29/:

$$\Xi(f) = \frac{\gamma}{v^2}\frac{\partial}{\partial v} v^3 \nu_{tr}(v) f^{(0}(v)$$

where $\gamma = m/M$, ν_{tr} is the transport collision frequency for electron-atom collisions.

Inelastic collisions are accounted for by the inelastic collision integral $\Xi^*(f)$, which consists of two contributions: the inelastic scattering of fast electrons on atoms, described by the term $\nu^*(v)f(\vec{v})$ (ν^* is the velocity dependent inelastic collision frequency), and the increase of the number of the slow electrons resulting from the inelastic collisions and ionization of the atomic particles. The second process is not essential for the ensemble of fast electrons responsible for the optical emission of an ionized gas and may be omitted in the spectropolarimetic studies. Taking into consideration that the Fokker-Plank term is isotropic, the zero-order multipole moment of the collision integral is given by the formula:

$$\Xi^{(0)}(f) = \frac{\gamma}{v^2}\frac{\partial}{\partial v} v^3 \nu_{tr}(v) f^{(0}(v) - \nu^* f^{(0)}(v) Z$$

Neither the contribution of the Fokker-Plank term nor the inelastic collisions into the multipole moments of the higher ranks are sufficient. Therefore for $\kappa \geq 1$, using (4.16) and (3.20), we have:

$$\Xi^{(\kappa)} = \frac{2\kappa+1}{4\pi}\nu(v)\int d\vec{n}\, C^{(\kappa)^*}(\vec{n})\left[\overline{f(v,\vec{n})} - f(v,\vec{n})\right] =$$

$$-\nu(v) f^{(\kappa)}(v,\vec{n}) + \frac{2\kappa+1}{4\pi}\frac{\nu(v)}{\sigma_{el}}\sum_{\kappa'} f^{(x')}(v)\int d\vec{n'}\, C_0^{(x)^*}(n/prime)\sigma_{el}(\vec{n},\vec{n'})$$

where

$$\sigma_{el} = \int d\vec{n}\,\sigma_{el}(\vec{n},\vec{n'})$$

is the total elastic cross-section and $\nu(v) = N_a v \sigma_{el}$ is the frequency of elastic collisions.

Analysing the last integral, let us calculate the spherical function $C_0^{(\kappa')}(n')$ within the reference frame of the quantization axis directed along $\vec{n}$ vector. Using the general rules of transformation of the basic vectors of the rotation group of symmetry and taking into consideration the properties of the components of D matrix, we have:

$$C_0^{(x')}(\vec{n'}) = \sum_q D_{q0}^{(\kappa')}(OY\vec{n})\, {}^{\vec{n}}C_q^{(\kappa')}(\vec{n'}) = \sum_q C_q^{(x')}(\vec{n})\, {}^{\vec{n}}C_q^{(\kappa')}(\vec{n'})$$

The axial symmetry of the scattering cross-section with respect to $\vec{n}$ vector gives the following relations:

$$\sigma_{el}(\vec{n},\vec{n'}) = \sigma_{el}(\beta); \quad \int d\vec{n'}\, {}^{\vec{n}}C_q^{(x')'}(\vec{n'})\sigma_{el}(\vec{n},\vec{n'}) = \delta_{0q}\sigma_{el}^{(\kappa)},$$

where $\beta = (\widehat{\vec{n},\vec{n'}})$ and

$$\sigma_{el}^{(\kappa)} = \int d\beta\, P_\kappa(\beta)\sigma_{el}(\beta).$$

Therefore we may write:

$$\Xi^{(\kappa)}(f) = -\nu(v)f^{(\kappa)}(v) + \frac{2\kappa+1}{4\pi}\frac{\nu(v)}{\sigma_{el}}\sum_{\kappa'}\int d\vec{n}\, C_0^{(\kappa)^*}(\vec{n})C_0^{(x')}(\vec{n})\sigma_{el}^{(\kappa)}$$

and finally, using the orthogonality of the spherical functions, we have:

$$\Xi^{(\kappa)}(f) = -\nu_\kappa(v)f^{(\kappa)}(v) \ (\kappa \geq 1),$$

where

$$\nu_\kappa(v) = \nu(v)[\sigma_{el} - \sigma_{el}^{(\kappa)}]/\sigma_{el}$$

is the collision frequency of the rank κ, $\nu_1(v) = \nu_{tr}(v)$.

Making use of the known relations for the Legendre polynomials /16/:

$$\alpha P_\kappa(\alpha) = \kappa/(\kappa+1)P_{\kappa-1}(\alpha) + (\kappa+1)/(2\kappa+1)P_{\kappa+1}(\alpha)$$

$$(1-\alpha^2)\partial P_\kappa(\alpha)/\partial\alpha = \kappa(\kappa+1)/(2\kappa+1)\left[P_{\kappa-1}(\alpha) - P_{\kappa+1}(\alpha)\right],$$

the left hand side of the equation (4.15) may be transformed. Finally, for the multipole moments of the different rank, we have the following set of equations:

$$\begin{aligned} &\frac{\partial f^{(0)}}{\partial t} + \frac{a}{3}\left(\frac{\partial}{\partial v} + \frac{2}{v}\right) f^{(1)} = \frac{\gamma}{v^2}\frac{\partial}{\partial v}v^3\nu_1(v)f^{(0)}(v) - \nu^*(v)f^{(0)}(v) \\ &\frac{\partial f^{(\kappa)}}{\partial t} + \frac{\kappa}{2\kappa-1}a\left(\frac{\partial}{\partial v} + \frac{\kappa+2}{v}\right) f^{(\kappa+1)} = -\nu_\kappa f^{(\kappa)}. \end{aligned} \tag{4.17}$$

For the trapped bulk plasma electrons with a low degree of anisotropy: $f^{(\kappa+1)} \ll f^{(\kappa-1)}$. Consequently, this equation may be written in the form:

$$\frac{\partial f^x}{\partial t} + \frac{\kappa}{2\kappa-1}a\left(\frac{\partial}{\partial v} - \frac{\kappa-1}{v}\right) f^{(\kappa-1)} = -\nu_\kappa f^{(\kappa)}.$$

For the first and second multipole moments we have the following equations:

$$\begin{aligned} \frac{\partial f^{(1)}}{\partial t} + a\frac{\partial f^{(0)}}{\partial v} &= -\nu_1 f^{(1)} \\ \frac{\partial f^{(2)}}{\partial t} + \nu_2 f^{(2)} = \frac{2}{3}\left[\frac{\partial}{\partial v} - \frac{1}{v}\right] af^{(1)} &= \frac{2}{3}v\frac{\partial}{\partial v}\frac{1}{v}af^{(1)} \end{aligned}$$

and the mean value of the quadrupole moment will be given by:

$$\langle f^{(2)}(v)\rangle = \frac{2}{3\nu_2}v\frac{\partial}{\partial v}\frac{1}{v}\langle af^{(1)}(v)\rangle,$$

where the corner brackets stand for the time averaging.

So, the mean value of the quadrupole moment of the velocity distribution function is connected to the product of the first multipole moment by the longitudial electric field strength. The power transmitted to the plasma per unit volume is determined by the same value. Indeed, the energy delivered per second to an electron from the electric field is equal to $e(\vec{E}\cdot\vec{v})$. Therefore the power transmitted per unit volume of the plasma and averaged over a oscillatoring field period is given by:

$$\langle W\rangle = \int d\vec{v}\, e\langle f(\vec{v})(\vec{E}\cdot\vec{v})\rangle = \int_0^\infty dv\, w(v),$$

where

$$\langle w(v)\rangle dv = mv^3\langle af^{(1)}(v)\rangle dv$$

is the power absorbed by electrons with the velocities in the range $v, v+dv$. Comparing the two expressions, we may write:

$$\langle f^{(2)}(v)\rangle = \frac{2}{3}\frac{v}{\nu_2(v)}\frac{\partial}{\partial v}\frac{\langle w(v)\rangle}{mv^4}.$$

Then the expression for $\langle w(v)\rangle$ will take the form:

$$\langle w(v)\rangle = \frac{3}{2}mv^4\int_v^\infty dv'\,\frac{\nu_2(v')}{v'}\langle f^{(2)}(v')\rangle$$

with the boundary conditions: $\langle w(0)\rangle = \langle w(\infty)\rangle = 0$.

In order to calculate the polarization degree we should obtain the expression for the quadrupole moment and its connection to the kinetic parameters of the electron gas. Let us analyse the temporal dependence of the multipole moments, using their expansion in Fourier series:

$$\begin{aligned} f^{(\kappa)}(v,t) &= \textstyle\sum_{n=-\infty}^{\infty} f_n^{(\kappa)} e^{in\omega_\sim t} \\ a(t) &= \tfrac{1}{2}a_\sim(e^{i\omega_\sim t} + e^{-i\omega_\sim t}) \end{aligned} \qquad (4.18)$$

where the subscript indicates the temporal harmonic.

The generalized system of equations for the harmonics of the multipole moments of the velocity distribution function will take the form:

$$in\omega_\sim f_n^{(0} + \frac{1}{6}\frac{a_\sim}{v^2}\frac{\partial}{\partial v}v^2(f_{n+1}^{(1)} + f_{n-1}^{(1)}) = \frac{\gamma}{v^2}\frac{\partial}{\partial v}v^3\nu_1 f_n^{(0)} - \nu^* f_n^{(0)}$$

$$in\omega_\sim f_n^{(\kappa)} + \frac{\kappa}{2\kappa+1}\frac{a_\sim}{2}\left(\frac{\partial}{\partial v} - \frac{\kappa-1}{v}\right)\left(f_{n-1}^{(\kappa-1)} + f_{n+1}^{(\kappa-1)}\right) = -\nu_\kappa f_n^{(\kappa)} \quad (\kappa\geq 1)$$

Assuming that the alternative field is relatively weak, the magnitude of harmonics with increasing numbers n is decreasing, so in the majority of cases it is possible to restrict oneself to the low n-numbers harmonics.

Using Fourier representation (4.18), we have:

$$\langle af^{(1)}(v)\rangle = \frac{1}{2}a_\sim(f_1^{(1)} + f_{-1}^{(1)})$$

and the equations for $f_\pm^{(1)}$ harmonics will take the form:

$$\begin{aligned} i\omega_\sim f_1^{(1)} + \nu_1 f_1^{(1)} &= -\tfrac{1}{2}\tfrac{\partial}{\partial v}a_\sim(f_0^{(0)} + f_2(0)) \\ -i\omega_\sim f_{-1}^{(1)} + \nu_1 f_{-1}^{(1)} &= -\tfrac{1}{2}\tfrac{\partial}{\partial v}a_\sim(f_0^{(0)} + f_{-2}(0)) \end{aligned}$$

Neglecting $f_\pm^{(0)}$ in comparison with $f_0^{(0)}$ and resolving this system, we have

$$\langle af^{(1)}(v)\rangle = -\frac{1}{2}\frac{\nu_1(v)a_\sim^2}{\omega_\sim^2 + \nu_1^2(v)}\frac{\partial}{\partial v}f_0^{(0)}(v) \tag{4.19}$$

and for the quadrupole moment we may write:

$$\langle f^{(2)}(v)\rangle = -\frac{a_\sim^2}{3\nu_2(v)}v\frac{\partial}{\partial v}\frac{1}{v}\frac{\nu_1(v)}{\omega_\sim^2 + \nu_1^2(v)}\frac{\partial}{\partial v}\langle f^{(0)}\rangle$$

This expression is formally the same as for the very simple model of the oscillative drift of an electron gas, if the collision frequency does not depend on the velocity of electron and if the mean square of the drift velocity is equal to:

$$\langle \Delta v^2\rangle = \frac{a_\sim^2\nu_1}{\nu_2(\omega_\sim^2 + \nu_1^2)}.$$

The degree of polarization of the studied spectral line for the low anisotropic axially symmetric plasma of the central part of the capacitive discharge may be written as:

$$P = -\frac{a_\sim^2}{5\sqrt{6}}w(JJ_1)\frac{\Gamma_0}{\Gamma_2}\frac{\int_{v_t}^\infty dv\,\sigma^{(2)}\frac{v^4}{\nu_2}\frac{\partial}{\partial v}\frac{1}{v}\frac{\nu_1}{\omega_\sim^2+\nu_1^2}\frac{\partial}{\partial v}\langle f^{(0)}\rangle}{\int_{v_t}^\infty dv\,\sigma^{(0)}v^3\langle f^{(0)}\rangle} \tag{4.20}$$

So, by measuring the degree of polarization in the center of the discharge for a known $f^{(0)}$, it is possible to determine the alternative electric field strength.

In order to determine the energy distribution function of the fast electrons responsible for the excitation of the emission spectrum, we use the time averaged equation (4.17). Substituting (4.19) we have:

$$\frac{1}{v^2}\frac{\partial}{\partial v}v^2\left[\frac{1}{6}\frac{\nu_1(v)a_\sim^2}{\omega_\sim^2 + \nu_1^2(v)}\frac{\partial}{\partial v} + \gamma v\nu_1(v)\right]\langle f^{(0)}(v)\rangle = \nu^*\langle f^{(0)}(v)\rangle.$$

Using energy as a variable instead of velocity we have:

$$\frac{\partial}{\partial\varepsilon}\left[D(\varepsilon)\frac{\partial}{\partial\varepsilon}+V(\varepsilon)\right]\langle f^{(0)}(\varepsilon)\rangle = u(\varepsilon)\langle f^{(0)}(\varepsilon)\rangle,$$

where

$$\begin{aligned} D(\varepsilon) &= \tfrac{1}{3}a_{\sim}^2 m\nu_1(\varepsilon)\varepsilon^{3/2}/(\omega_{\sim}^2+\nu_1^2(\varepsilon)) \\ V(\varepsilon) &= 2\gamma\nu_1(\varepsilon)\varepsilon^{3/2}; \qquad\qquad u(\varepsilon)=\nu^*(\varepsilon)\varepsilon^{3/2}. \end{aligned}$$

Let us seek a solution to this equation in the form:

$$\langle f^{(0)}(\varepsilon)\rangle = A\exp\left[\int_{\varepsilon_t}^{\varepsilon} d\varepsilon'\,\eta(\varepsilon')\right]$$

where A is a numerical constant and $\eta^{-1}(\varepsilon)$ characterizes the scale of the exponential decrease of the distribution function, or the local temperature of electrons with energy around ε .

For $\eta(\varepsilon)$ function we have the equation:

$$D(\varepsilon)\frac{\partial\eta}{\partial\varepsilon}+D(\varepsilon)\eta^2+\left[\frac{\partial D(\varepsilon)}{\partial\varepsilon}+V(\varepsilon)\right]\eta+\frac{\partial V(\varepsilon)}{\partial\varepsilon}-u(\varepsilon)=0,$$

which was numerically solved, beginning from high energy ε_a and approaching ε_t step by step, using the standard Runge-Kutta procedure. The accuracy and the stability of the approximate numerical solution were tested by increasing ε_a and decreasing h parameters.

The spectropolarimetric determination of the amplitude of the alternative high frequency field for the central part of the discharge plasma was performed in an helium discharge maintained in a cylindric cell. Spectropolarimetric data on the singlet 492.2 nm helium line were used at 155 mTorr, in order to reduce the effect of the loss cones forming near the side walls. The mean free path of electrons was 8 mm, much lower than the diameter of the discharge cell.

Using (4.20) and the expression of the electron impact alignment cross-section in terms of $P_0(\varepsilon)$ function (1.53), the second Stokes parameter, takes the form:

$$P = -\frac{2}{15}\frac{e^2E_{\sim}^2}{m}\frac{\Gamma_0}{\Gamma_2}\frac{\int_{\varepsilon_t}^{\infty} d\varepsilon\,\varepsilon^2\sigma(\varepsilon)\left[P_0^{-1}(\varepsilon)-1/3\right]^{-1}\nu_2^{-1}(\varepsilon)\frac{\partial}{\partial\varepsilon}\left[\frac{\nu_1(\varepsilon)\eta(\varepsilon)}{\omega_{\sim}^2-\nu_1^2(\varepsilon)}\langle f^{(0)}(\varepsilon)\rangle\right]}{\int_{\varepsilon_t}^{\infty} d\varepsilon\,\varepsilon\sigma(\varepsilon)\langle f^{(0)}(\varepsilon)\rangle} \tag{4.21}$$

where $\langle f^{(0)}(\varepsilon)\rangle$ is determined by expression (4.33).

The functions $\nu_{1,2}(\varepsilon)$ and $\nu^*(\varepsilon)$ were obtained on the basis of data presented by Drukarev /32/. The ratio Γ_0/Γ_2 was experimentally determined

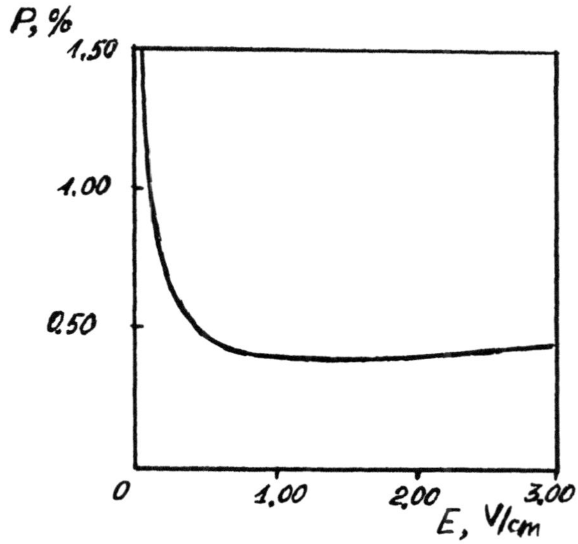

Figure 58. Calculated degree of polarization ofthe 492.2 nm spectral line as a function of the amplitude of the alternative electric field in a high frequency capacitive discharge

by the measurement of the pressure dependence of the Hanle signal width. For 155 mTorr, $\Gamma_0/\Gamma_2 = 0.33(3)$. The calculations of the $\eta(\varepsilon)$ function and of the integrals in (4.21) were made numerically. The calculation of the degree of polarization, as a function of the amplitude of the high frequency alternative field (Fig. 58), showed a decrease in the polarization degree on the first stage with the growth of the high frequency field amplitude. This effect was explained as due to the effective heating of the electron gas. The increase in the chaotic thermal motions was the reason for the decrease of the ratio $f^{(2)}/f^{(0)}$. For an electric field exceeding 1 V/cm, the degree of polarization approached 0.4% and did not depend on the electric field amplitude. After that an increase in the polarization degree took place. The amplitude of the alternative electric field strength for the conditions of the spectropolarimetric experiments on 492.2 nm helium line ($V_\sim = 30V$, $\omega_\sim = 100$ MHz, $p = 155$ mTorr) in the center of a capacitive high frequency discharge, of length 6 cm and diameter 4 cm, was found to be (0.35 ± 0.05) V/cm.

CHAPTER 5

POLARIZATION SPECTROSCOPY OF IONS

Spectropolarimetric effects in the linear spectrum of the optical emission are connected not only to the anisotropy of the fast exciting electrons or to the distribution of the resonance radiation but also to the anisotropy in the relative velocity space of the colliding atomic particles. In the last case, the anisotropic collisional relaxation within the system of the quantum states of the excited atomic particles plays an important role. Such anisotropic relaxation takes place under the collisional relaxation of the drifting ions with the neutral atomic particles. The present chapter is devoted to the description of the drift self-alignment of ions and the diagnostic applications of this effect for the ionized gas sensing.

5.1 Theory of drift self-alignment

In the first chapters, the collisional relaxation superoperator $\hat{\hat{\Gamma}}$ was assumed to be isotropic. Now we shall examine the polarization effects resulting from the anisotropy of the relaxation processes.

5.1.1 THEORY

Let us examine now the different principal self-alignment mechanisms resulting from the anisotropy of the motion of the atomic particles in an ionized gas. This type of self-alignment may be expected within anisotropic ionized gas entities but also sometimes in simple gas discharges at low pressure. The most simple example is the superposition of isotropic motions with a Maxwellian velocity distribution with a drift in an external electric field $E(\vec{r})$ with a velocity $v = \beta' E(\vec{r})$ (β' is the ion mobility). In case of a drift $v_0 = v_0 e_z$ along the OZ axis, the distribution function of ions relative velocities with respect to atoms $\vec{v} = \vec{v_a} - \vec{v_1}$ may be expressed by:

$$f(v) = (a/\pi)^{3/2} \exp\left[-a(v^2 + v_0^2 + 2vv_0 \cos\theta)\right]$$

Where θ is the angle between the drift velocity $\vec{v_0}$ and the relative velocity vector $\vec{v}$ and a is a parameter expressed in terms of masses (m_a, m_i) and temperatures (T_a, T_i) of atoms and ions:

$$a = \frac{m_i m_a}{2k(m_i T_a + m_a T_i)}.$$

The degree of anisotropy of the relative ions-neutrals velocities is characterized by the dimensionless parameter $\lambda = v_0\sqrt{a}$ that, for $\lambda = 0$, corresponds to the isotropic distribution and for $\lambda \to \infty$ gives the three-dimensional δ function $\delta(\vec{v} - \vec{v_0})$.

The drift of ions in an ionized gas brings about a preferential direction within a plasma, so the probability of collisions along the drift velocity and in the perpendicular direction have different values. Such partial anisotropy of the collisional interaction of ions may be the reason for the anisotropic collisional relaxation, within the system, of the close fine structure levels of the ion energy structure. According to the general theory of the anisotropic collisional relaxation of the density matrix /75, 76/, the polarization moments of the ensemble of ions are described by the kinetic equation:

$$\frac{d}{dt}\rho_0^{(\kappa)}(J) = \sum_{\kappa_1} R_0^{\kappa\kappa_1}(JJ_1)\rho_0^{(\kappa_1)}(J_1)$$

where $R_0^{\kappa\kappa_1}(JJ_1)$ is the matrix of the partial anisotropic collisional relaxation of the longitudinal polarization moments $\rho_0^{(\kappa_1)}(J)$, which is expressed in terms of the total anisotropic collisional relaxation matrix $\tilde{R}_Q^{\kappa\kappa_1}(JJ_1v)$:

$$R_0^{\kappa\kappa_1}(JJ_1) = \frac{4a^{3/2}}{\sqrt{\pi}}e^{-av^2}\sum_{Q,l}(-1)^Q \begin{bmatrix} \kappa & \kappa_1 & l \\ 0 & 0 & 0 \end{bmatrix} \begin{bmatrix} \kappa & \kappa_1 & l \\ Q & -Q & 0 \end{bmatrix} \cdot$$

$$\int dv\, \tilde{R}_Q^{\kappa\kappa_1}(JJ_1v)v^2 e^{-av^2}\left(\frac{v}{2av_0}\right)^l \left(\frac{1}{v}\frac{d}{dv}\right)^l \frac{\sinh 2avv_0}{2avv_0}$$

In the case of anisotropic relaxation, the projection of the momentum on the drift axis is preserved, but not the rank of the tensor κ. Therefore, due to the linkage of momenta with different ranks, the drift of ions is accompanied by the formation of a longitudinal alignment with the axis parallel to the drift velocity vector. This process is described by this equation for $\kappa_1 = 0,\ q = 0,\ \kappa = 2$.

The study of drift self-alignment of ions accounting for the interaction of colliding particles, that incorporates polarization interaction of the charged ion with the neutral atom, dispersive interaction and combined polarization interaction of the neutral atom with the charge and the quadrupole moment of the ion, reveals a number of specific features of this phenomenon. It was shown that a general requirement for the formation of drift self-alignment is the deviation of the fine level population from its equilibrium values. Efficiency of drift self-alignment is dependent on the degree of anisotropy of motion and of the energy of an ion. This process is only possible if the kinetic energy exceeds the multiplet splitting $m_1v_1^2/2 > \Delta E$. Under ordinary low temperature plasma conditions /77, 78/, the last condition may be fulfilled for a number of excited states of ions.

5.1.2 CHARACTERISTIC RATE CONSTANTS

Let us take as a typical example the formation of drift self-alignment within the array of fine structure substates of a noble gase atom moving in an electric field in an atmosphere of neutral particles of the same gas. For the drift self-alignment of the narrow $^2P_{1/2,3/2}$ noble gas doublet under collisions with noble gas atoms in the ground state $^1S_0$, the equations for the population $n_J = \sqrt{2J+1}\,\rho^{(0)}(J)$ and longitudinal alignment

$$a_J = \frac{\langle 3J_z^2 - J^2 \rangle}{J(2J+1)} = \sqrt{\frac{(J+1)(2J+1)(2J+3)}{5J(2J-1)}}\rho_0^{(2)}(J)$$

will take the form :

$$\begin{aligned} \dot{n}_{1/2} &= -W_{tr}n_{1/2} - \tfrac{1}{2}W_{tr}n_{3/2} + \tfrac{1}{2}W_a a_{3/2} \\ \dot{n}_{3/2} &= W_{tr}n_{1/2} - \tfrac{1}{2}W_{tr}n_{3/2} - \tfrac{1}{2}W_a a_{3/2} \\ \dot{a}_{3/2} &= W_a n_{1/2} - \tfrac{1}{2}W_a n_{3/2} - \tfrac{1}{2}W_d a_{3/2}, \end{aligned}$$

where

$$W_{tr} = \sqrt{2}R_0^{00}(\frac{1}{2}\frac{3}{2}) = -R_0^{00}(\frac{1}{2}\frac{1}{2}) = -2R_0^{00}(\frac{3}{2}\frac{3}{2}) = \sqrt{2}R_0^{00}(\frac{3}{2}\frac{1}{2}) = N\langle v\sigma_{tr}\rangle$$

is the probability of the intramultiplet collisional population transfer with the rate constant $\langle v\sigma_{tr}\rangle$,

$$W_a = \sqrt{2}R_0^{20}(\frac{3}{2}\frac{1}{2}) = \sqrt{2}R_0^{02}(\frac{1}{2}\frac{3}{2}) = -2R_0^{20}(\frac{3}{2}\frac{3}{2}) = -2R_0^{02}(\frac{3}{2}\frac{3}{2}) = N\langle v\sigma_a\rangle$$

is the probability of the drift self-alignment $a_{3/2}$, formed from the populations of doublet components with rate constant $\langle v\sigma_a\rangle$, and

$$W_d = -R_0^{22}(\frac{3}{2}\frac{3}{2}) = N\langle v\sigma_d\rangle$$

is the probability of the collisional depolarization of the state $J = 3/2$ with rate constant $\langle v\sigma_d\rangle$.

Taking into account for the stationary isotropic excitation of the fine structure levels

$$F_{1/2} = \Gamma_0 n_{1/2}(p=0),\ F_{3/2} = \Gamma_0 n_{3/2}(p=0)$$

and radiative decay we have :

$$\begin{aligned} -W_{tr}B + \tfrac{1}{2}W_a a_{3/2} - \Gamma_0 n_{1/2} + F_{1/2} &= 0 \\ W_{tr}B - \tfrac{1}{2}W_a a_{3/2} - \Gamma_0 n_{1/2} + F_{3/2} &= 0 \\ W_a B - W_d a_{3/2} - \Gamma_0 a_{3/2} &= 0 \end{aligned} \tag{5.1}$$

where $B = n_{1/2} - n_{3/2}/2$.

Using the last equation, we have :

$$A(p) = \frac{a_{3/2}}{n_{1/2} - n_{3/2}/2} = \frac{N(v\sigma_d)}{N\langle v\sigma_d\rangle + \Gamma_0} \tag{5.2}$$

At low pressures, $A(p)$ is a linear function of pressure:

$$A(p) = p\langle v\sigma_a\rangle/\Gamma_0 kT.$$

$A(p)$ is expressed in terms of only one rate constant $\langle v\sigma_a\rangle$. For higher pressures ($\Gamma_0 \ll N\langle v\sigma_a\rangle$), $A(p)$ tends to the value $\langle v\sigma_a\rangle/\langle v\sigma_d\rangle$. It is therefore possible to determine the alignment and depolarization rate constants by making use of the $A(p)$ function for a wide range of pressures.

For the rate constant of the collisional population transfer between the fine structure levels, using the first two equations of system (5.1) we have the following expression :

$$\langle v\sigma_{tr}\rangle = \frac{2}{3}\frac{\Gamma_0}{N}\left[\frac{B(0)}{B(p)} + \frac{3}{4}\frac{N}{\Gamma_0}\langle v\sigma_a\rangle A(p) - 1\right]$$

The characteristic rate constants describing the drift self-alignment in an ionized gas may be determined on the basis of the experimental data for $A(p)$ and $B(p)$ dependencies. The inverse problem is straightforward. This technique enables one to measure the drift velocity of ions in an ionized gas using the experimental spectropolarimetric data and the calculation of the relaxation matrix elements for a given pair of colliding particles.

5.2 Observation of drift self-alignment in a hollow cathode

Observation of drift self-alignment implies the solution of a number of specific experimental problems: the choice of an ionized gas entity with a high degree of anisotropy of ion motions and the separation of the different mechanisms that could lead to the polarization of the excited ionic states.

5.2.1 SPECTROPOLARIMETRIC PECULIARITIES

Anisotropic collisional relaxation may be expected only in a discharge plasma with a high degree of anisotropy of motion of ions. In the positive column plasma of a d.c. discharge and in a high frequency discharge, the drift of ions is rather slow. Ion velocities may be considerable only within the thin boundary sheath, but the spectroscopic observation of this region is rather complicated. The opposite situation takes place in a hollow cathode. It is known that the potential drop within the hollow cathode occurs

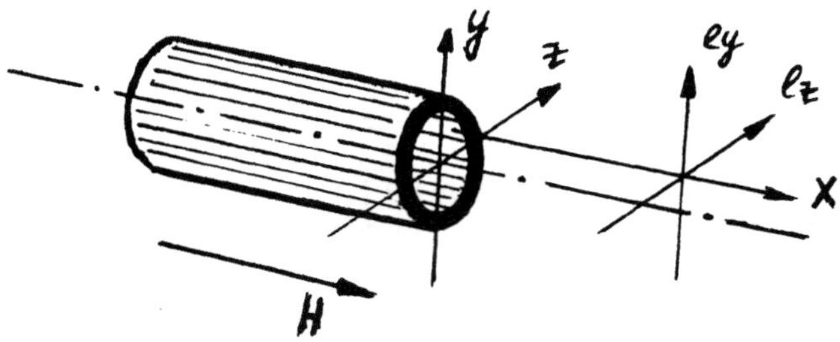

Figure 59. Experimental scheme for observation of the spectropolarimetric effects on ionic levels of noble gases in a hollow cathode discharge

mainly in the dark cathode space with a width up to 1 - 2 millimeters for the discharge in noble gases in a cathode of 1 - 2 cm diameter and pressure lower than 1 Torr. The main volume of the hollow cathode is filled by the negative glow discharge plasma.

The distribution of electric field within the hollow cathode is rather nonuniform. Large in the dark space, the electric field decreases rapidly while moving to the discharge center. In view of this, the drift of ions in the boundary region of the negative glow must be very effective. Mass spectroscopic measurements confirm that practically all positive ions are formed in the negative glow and then drift to the wall of the hollow cathode. The energy of ions at the wall surface is determined by the potential drop in the discharge volume. Observations of the spectropolarimetric effects were realized in a cylindric alumimium hollow cathode of 1 cm bore and 5 cm length in the direction of its symmetry axis (Fig. 59), for a discharge voltage 80 - 100 V /79, 80/. The quantity $(I_z - I_y)_x$, which is proportional to the alignment of the angular momenta, was measured for atomic and ionic lines by means of a magnetic spectropolarimeter. Some anomalies have been detected in the polarization of ionic spectral lines, corresponding to the transitions from closely separated levels of the fine structure of the singly charged Ar ion (Fig. 60) within the pressure range 0.1 - 0.4 Torr and for a discharge current 10 - 60 mA. A number of typical control experiments confirmed that these magnetic polarization signals have been determined by the destruction of the self-alignment of the ionic states by the magnetic field. The signs of signals were opposite for the spectral lines on transitions from the same upper level to lower states with $\Delta J = 1$. No linear polarization was observed on the line from the upper level with $J = 1/2(4d'^2P_{1/2}, \lambda = 375.5\text{nm})$, which cannot be aligned, but strong linear polarization was detected from the second component of this doublet with $J = 3/2(4d'\,^2P_{3/2}, \lambda = 381.9$ nm). The shape of the magnetic depolarization signals exhibited the characteristic dependence on the position of the

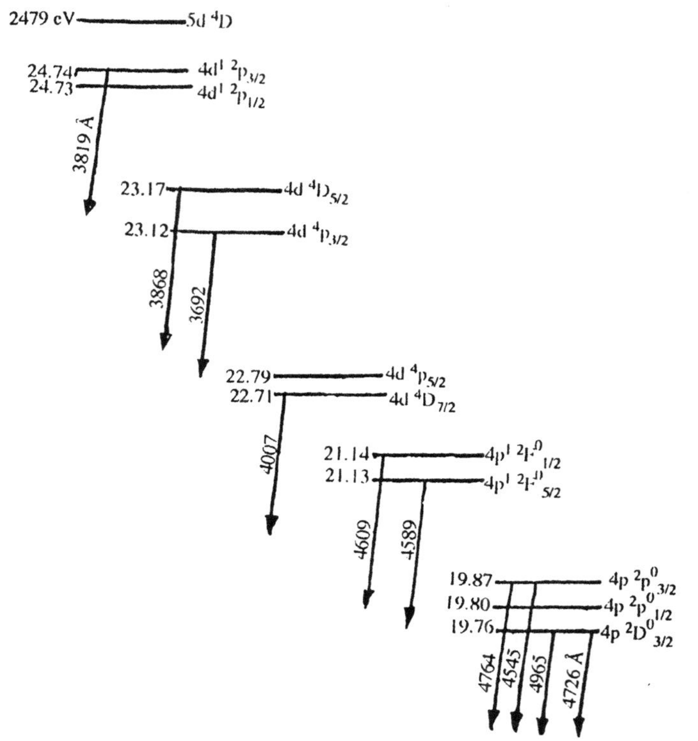

Figure 60. **Scheme of the levels of the energy structure of ArII showing the spectral transitions on which drift self-alignment has been detected**

studied volume, reflecting the radial symmetry of the self-alignment. The magnitude of the signal was maximum at the boundary region of the negative glow for the different observation conditions. All these experimental data displayed that the ionic states of argon are effectively aligned and this self-alignment is radially distributed in the negative glow discharge plasma of the cylindrical hollow cathode.

5.2.2 EVIDENCE FOR DRIFT SELF-ALIGNMENT

A number of special experiments revealed the physical mechanism of the self-alignment of argon ion multiplets /79, 80/. The first principal result was that the radial dependence of the self-alignment of ionic states was non-monotonous and differed crucially from that of the intensity $I_0 = I_z + I_y$ of the same spectral line (Fig. 61). The maximum polarization degree was near the boundary of the negative glow and the intensity maximum was located in the center of the negative glow. The radial intensity profile is explicable on the basis that the excitation of the neutral particles, and the

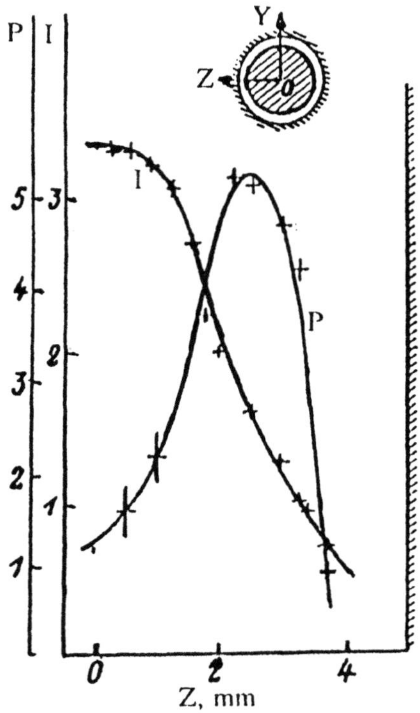

Figure 61. **Radial dependence of the first (I) and second (P) Stokes parameters of ArII spectral line 381.9 nm emitted by the hollow cathode**

formation of the spectral characteristics of emission of the ionized gas in the hollow cathode discharge at low pressure, comes mainly from collisions with electrons. The distribution function of electrons in the hollow cathode is essentially anisotropic. The primary electrons, appearing due to cathode surface bombardment by ions, are effectively accelerated to the center of the discharge within the dark space. The density of the fast electrons and the probability of excitation are maximum in the central part of the hollow cathode. This is well-represented by the characteristic bell shape of the intensity profile. The polarization profile has a different shape. The principal result allowing to eliminate electron impact self-alignment as the explanation of boundary spectropolarimetric effects was the study of the pressure dependence of the magnetic contour widths. If self-alignment is determined by electron impact, the extrapolation of the Hanle signal widths must give the radiative lifetime of the upper level of the spectral transition. For all the transitions of the studied argon ions, extrapolation to zero pressure gave effective lifetime values significantly different from the radiative lifetimes known from independent reliable measurements. Another important experimental result in this respect was the comparison of the polarization effects on atomic spectral lines, like the transitions from 5p atomic states of argon, with the effects on lines from ions excited in the same discharge. It is known that direct excitation is realized by fast beam electrons whose mean energy exceeds the thermal bulk electron energy (10 eV for 0.3 Torr and 20 - 30 mA). Despite that the maxima of the electron impact cross-sections of 5p levels (14.5 - 14.7 eV) is much closer to the mean electron beam energy than that for the studied argon ion levels (35.5 - 40.5 eV), the degree of self-alignment of 5p states in the boundary region of the negative glow was much less (one order of magnitude less) than that for the ionic lines. The low degree of self-alignment of 5p atomic levels was not connected to the low magnitude of the electron impact alignment cross-section. Indeed, strong polarization of the spectral lines originating from 5p argon levels has been detected in the positive column of the d.c. discharge maintained in a 6 mm bore tube for the same pressure and current density range. It was determined by the formation of loss cones in the velocity distribution of the fast electrons (see section 3.3.1). So, the low degree of self-alignment of 5p states in the hollow cathode indicates only the comparatively low degree of direct electron impact excitation anisotropy of these levels. On the other hand, comparing the excitation threshold of the ionic levels studied with 5p levels and the mean energy of the fast beam electrons, it is evident that the anisotropy of the exciting threshold electrons producing the direct excitation of ionic levels must be lower than that for 5p levels. Therefore, if the alignment of the atomic particles is determined by direct electron impact excitation, the degree of linear polarization of the spectral lines originating from 5p atomic levels must be much higher than that for the ionic states. Experiments gave the opposite result.

All these supplementary experiments allowed us to conclude that the physical reason for anamolous self-alignment of ion levels in the hollow cathode is the drift of ions. Energy considerations supported the possibility of such process. The electric field for which an argon ion may have a kinetic energy of the same order as intramultiplet splitting energy (0.01 - 0.1 eV) showed that at a pressure of 0.3 Torr the same kinetic energy is achieved in electric fields of 5-20 V/cm. Such electric fields really exist in the boundary of the negative glow and dark space of the hollow cathode discharge, where anamolous polarization effects have been observed. In ionized argon, the mean free path of an argon ion at 0.3 Torr is 0.2 mm, which is close to the radial dimension of the boundary intermediate region.

The basic feature of drift self-alignment is the complex nature of this effect. The observed polarization degree of ion lines results from a number of collisional processes with participation of the excited ions and normal atoms: anisotropic relaxation accompanied by the formation of the self-alignment intramultiplet collisional mixing of populations and depolarization. The last process leads to alignment destruction and results in the decrease of drift self-alignment efficiency. As it was shown by Petrashen, /76/, the total probability of self-alignment due to anisotropic collisional relaxation is a function of the population ratio of doublet ionic states. No drift self-alignment is induced, if the population distribution is thermal. Drift self-alignment is generated mainly due to the deviation of the relative population N_1/N_2 from the equilibrium value.

Taking into consideration the principal importance of the initial population inequality of the multiplet states in the production of the drift self-alignment of excited ions, the population ratio N_1/N_2 has been experimentally estimated for the same range of discharge parameters. The spectral line intensity method was used. It was shown that the population ratio of multiplets differs systematically from the ratio of the statistical weights of these levels g_1/g_2 .

Because of the complex nature of the self-alignment, we can assert with confidence that drift self-alignment must exist only within a narrow region of discharge conditions /79, 80/. Indeed, the spectropolarimetric effects on the ion states have been observed only within a narrow range of experimental parameters and appear to be very pressure selective.

In conclusion, all these experimental observations and test experiments allow one to advocate that the anamolous spectropolarimetric effects in the boundary region of the negative glow of the hollow cathode discharge plasma result from the drift self-alignment of components of the ionic multiplet structure.

5.3 Drift velocity of ions

The drift self-alignment effect provides new possibilities for the remote spectral sensing of ionized gases. Implementation of this technique is connected to the calculation of matrix elements of the anisotropic collisional relaxation matrix.

5.3.1 DETERMINATION OF THE RATE CONSTANTS

The potentialities of the drift self-alignment effect were demonstrated using a hollow cathode argon discharge /81/. The discharge was maintained in an aluminium hollow cathode of length 5 cm and diameter $2R = 1$ cm, at a range of pressures 0.01 - 0.5 Torr and discharge current 15 - 25 mA. Observation of the emission of the discharge plasma was carried out from different small parts of it, defined by z - coordinate, along the symmetry axis of the discharge. Two different doublets of the energy level system of the argon ion were used for the spectropolarimetric measurements: $4p\,^2P^o_{1/2,3/2}$ ($\Delta E = 0.07$) eV and $4d'\,^2P_{1/2,3/2}$ ($\Delta E = 0.01$ eV) on lines corresponding to transitions to one low level, $4s^2\,P_{3/2}$ and $4p^2D^o_{3/2}$ respectively. The radial dependencies of the first and second Stokes parameters showed that the maximal value of the first parameter I was observed in the center of the negative glow ($z = 0$) and the monotonous decrease to the boundary region was typical. At the same time, the maximum of the second Stokes parameter was located at the boundary of the negative glow.

Using (1,27), the intensity of a spectral line with polarization defined by the polarization vector $\vec{e}_\lambda$ may be expressed as:

$$I_{\vec{e}_\lambda} = C \mid \langle J_q \parallel d \parallel J \rangle \mid^2 (2J_q + 1)(-1)^{J+J_1} \cdot$$

$$\sum_{\kappa,q} (-1)^{\kappa+q} \sqrt{2\kappa+1}\, a^*_{q1} a_{q2} \begin{Bmatrix} J & J & \kappa \\ 1 & 1 & J_q \end{Bmatrix} \begin{bmatrix} \kappa & 1 & 1 \\ -q & q_2 & q_1 \end{bmatrix} \rho^{(\kappa)}_q(J)$$

where $a_{q1,q2}$ are the coefficients of the expansion, in a Cartesian system, of the polarization vector $\vec{e}_\lambda$ in terms of the circular unit vectors $\vec{e}_q (q = 0, \pm 1)$:

$$a_{\pm 1} = \pm i \sin\theta \cdot \exp(\pm i\varphi)/\sqrt{2},\ a_0 = -i\cos\theta.$$

For the frame of reference used in these experiments, the polarization vectors were oriented along the drift axis $\vec{e}_z$ and orthogonally $\vec{e}_y$. The coefficients for these two cases are expressed as ($a_{\pm 1} = 0$, $a_0 = -i$) and ($a_{\pm 1} = 1/\sqrt{2}$, $a_0 = 0$). Expressions for the linearly polarized components in the spectral line are given by:

$$\begin{aligned} I_z &= C \mid \langle J_1 \parallel d \parallel J \rangle \mid^2 (2J_1 + 1)(-1)^{J+J_1} \\ &\left[\begin{Bmatrix} J & J & 0 \\ 1 & 1 & J_1 \end{Bmatrix} \rho^{(0)}_0(J) - \sqrt{2} \begin{Bmatrix} J & J & 2 \\ 1 & 1 & J_1 \end{Bmatrix} \rho^{(2)}_0(J) \right] \end{aligned}$$

$$\begin{aligned} I_y &= C \mid \langle J_1 \parallel d \parallel J \rangle \mid^2 (2J_1 + 1)(-1)^{J+J_1} \\ &\left[\frac{1}{\sqrt{2}} \begin{Bmatrix} J & J & 2 \\ 1 & 1 & J_1 \end{Bmatrix} \rho^{(2)}_0(J) + \begin{Bmatrix} J & J & 0 \\ 1 & 1 & J_1 \end{Bmatrix} \rho^{(0)}_0(J) \right]. \end{aligned}$$

Calculating the reduced matrix element $\mid \langle J_1 \parallel d \parallel J \rangle \mid$ we obtain the relationship between the observables and the polarization moments:

$$\begin{aligned} I_y - I_z &= C \mid \langle J_1 \parallel d \parallel J \rangle \mid^2 (2L_1 + 1)(2J_1 + 1)(2J + 1) \cdot \\ &(-1)^{J+J_1} \frac{3}{\sqrt{2}} \begin{Bmatrix} J & J & 2 \\ 1 & 1 & J_1 \end{Bmatrix} \begin{Bmatrix} L_1 & J_1 & S \\ J & L & 1 \end{Bmatrix} \rho^{(2)}_0(J) \\ I_y + \tfrac{1}{2} I_z &= C \mid \langle J_1 \parallel d \parallel J \rangle \mid^2 (2L_1 + 1)(2J_1 + 1)(2J + 1) \cdot \\ &(-1)^{J+J_1} \frac{3}{2} \begin{Bmatrix} J & J & 0 \\ 1 & 1 & J_1 \end{Bmatrix} \begin{Bmatrix} L_1 & J_1 & S \\ J & L & 1 \end{Bmatrix} \rho^{(0)}_0(J) \end{aligned} \tag{5.3}$$

Neglecting the difference between the transition frequencies to the same lower level ($\Delta\lambda/\lambda \ll 1$) of the components of the doublet and expressing the quantities from formula (5.1) in terms of the spectropolarimetric parameters, we may write the formula for the A-function as follows:

$$\begin{aligned} A_{4d^1\,^2p} &= [12.5(I_y - I_z)_{3/2}]/[(I_y + \tfrac{1}{2}I_z)_{3/2} - (I_y + \tfrac{1}{2}I_z)_{1/2}] \\ A_{4d^2p^o} &= [(I_y - I_z)_{3/2}]/[(I_y + \tfrac{1}{2}I_z)_{3/2} - 2(I_y + \tfrac{1}{2}I_z)_{1/2}]. \end{aligned} \tag{5.4}$$

It is possible to express the rate constants of the collisional transfer of populations of the doublet components $\langle v\sigma_{tr}\rangle$ in terms of observable spectropolarimetric data as well. Using the general relation (5.3) for $4p^2P^o_{1/2,3/2}$ doublet we have:

$$\frac{B(0)}{B(p)} = \frac{\left[(I_y + \frac{1}{2}I_z)_{3/2} - 2(I_y - \frac{1}{2}I_z)_{1/2}\right]\,(p=0)}{\left[(I_y + \frac{1}{2}I_z)_{3/2} - 2(I_y - \frac{1}{2}I_z)_{1/2}\right]\,(p)}$$

Now it is possible to express the rate constant $\langle v\sigma_{tr}\rangle$ in terms of the observable spectropolarimetric parameters. At low pressure the function A is proportional to $\langle v\sigma_a\rangle p/kT_a\Gamma_0$ and at high pressure this function approaches $\langle v\sigma_a\rangle/\langle v\sigma_d\rangle$. The quantity $B(p)$ at low pressure is proportional to gas pressure as $\langle v\sigma_{tr}\rangle p/kT_a\Gamma_0$ and at high pressure this function tends to

$$\left[\langle v\sigma_{tr}\rangle - \langle v\sigma_a\rangle^2/2\langle v\sigma_d\rangle\right]\frac{p}{kT_a\Gamma_0}.$$

5.3.2 REMOTE SENSING OF THE HOLLOW CATHODE

Determination of the intensity $I_{y,z}(1/2, 3/2)$ of the linearly polarized components of spectral lines has been performed with the help of a magnetic polarization spectrometer. Using these spectropolarimetric data, the family of curves $A(p,\lambda)$ have been plotted (Fig. 62). For all cases, the $A(p,\lambda)$ function has an initial linear part and then goes to saturation. The radial dependence of this function reflects the radial variation of the drift velocity of the positive ions, which appear in the central part of the negative glow due to direct ionization, and then drift to the cathode wall. The strong radial dependence of the $A(p,\lambda)$ function indicates the high sensitivity of this method for drift velocity determination. The initial linear part of $A(p,\lambda)$ dependencies was used for estimating the rate constants of induction of self-alignment. At low pressure, the cross-section of this process has been approximated as $\langle v\sigma_a\rangle = kT_a\Gamma_0 A/p$. Using spectropolarimetric data for the

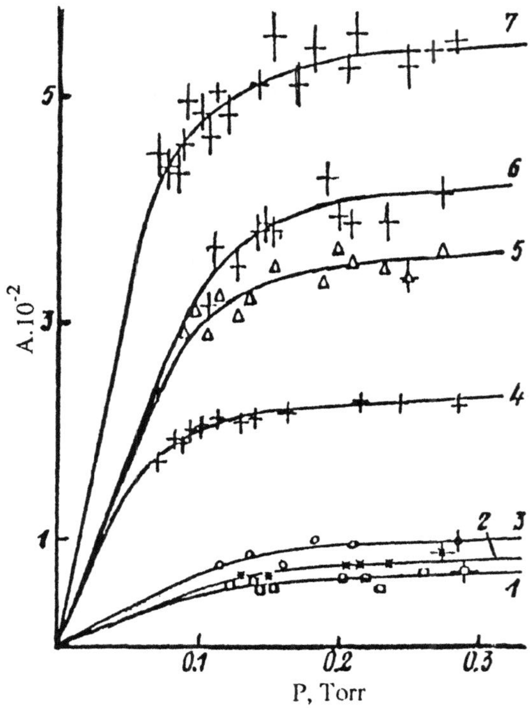

Figure 62. Pressure variation of the A(p) function for the $4d'P_{1/2,3/2}$ doublet of ArII ion in the hollow cathode at various radial distances z(mm):(1) - 0, (2) -0.5, (3) - 1, (4) - 3.3, (5) - 2.25, (6) - 2.5, (7) - 3.0

$4d^1\ ^2P$ doublet for different z-coordinates, corresponding to different drift velocities, we have:

$$\begin{aligned}
\langle v\sigma_d\rangle \cdot 10^{-10}\text{cm}^3\text{sec}^{-1} &= 3, & z &= 0-0.5\,\text{mm}\\
&= 4, & z &= 1\,\text{mm}\\
&= 20, & z &= 2.5\,\text{mm}\\
&= 40, & z &= 3.0\,\text{mm}
\end{aligned}$$

The quantity Γ_0 was taken from reliable direct measurements. The same estimates for $4p^2\ P_0^0$ doublet, using the linear part of $A(p)$ dependence for the polarization maximum, gave for the rate constant of the induction of collisional drift self-alignment at a low pressure range $\langle v\sigma_a\rangle = (7.5 \pm 2.0) \cdot 10^{-10}$ cm^3 sec^{-1}.

The experimental derived $B(p)$ dependence is represented by a linear function of the pressure up to 0.12 Torr. Rate constant for the collisional transfer of population was estimated from the slope of the linear dependence: $\langle v\sigma_{tr}\rangle = B(p)kT_a\Gamma_0/p = 1.8(8)10^{-9}$ cm^3 sec^{-1}. Pressure dependence must be considered for the analysis of the experimental data. The mean free path of ions decreases with pressure, as well as the mean drift velocity v_0 and the anisotropy parameter λ. As a result, the rate constants of the collisional processes will be pressure dependent. Such dependence must be the most pronounced for the rate constant of collisional self-alignment $\langle v\sigma_a \rangle$. For the isotropic case, $\lambda = 0$, this rate constant is null. Taking into consideration that the drift velocity is z-dependent, the rate constant must be coordinate-dependent as well. Pressure dependence of the rate constants of depolarization and collisional transfer of population are weaker. Even for an isotropic case, i.e. without any drift of ions ($\lambda = 0$), these constants are not zero. Such considerations are in agreement with the observed experimental dependencies for ArII doublets. Diagnostic possibilities of the polarization spectroscopy of an ion component are based on the relationship between rate constants, describing the drift self-alignment of ions, and the elements of the anisotropic collisional relaxation matrix of the polarization moments of the multiplet components $J = 1/2, 3/2$, /82/. Using expression (5.2) the function $A(p, \lambda)$ takes the form:

$$A(p, \lambda) = \frac{\sqrt{2} R_0^{02}(1/2, 3/2)}{\Gamma_0 - R_0^{22}(3/2, 3/2)}. \tag{5.5}$$

Elements of the matrix of anisotropic collisional relaxation may be represented as

$$R_q^{\kappa\kappa_1} = S_q^{\kappa\kappa_1}(\lambda)\beta_i N,$$

where $S_q^{\kappa\kappa_1}(\lambda)$ is function of the velocity distribution anisotropy of the colliding atomic particles and may be determined on a basis of calculations /76/. The quantity β_i depends on the anisotropy of polarization and of the dispersive ion-atom interaction:

$$\beta_i = |, (9Q_i e^2 \alpha_a/2 + \Delta)/\hbar\, |^{2/5} \gamma^{-3/10}$$

where $Q_i e^2$ is the quadrupole moment of the electron shell of the excited ion, α_a is the polarizibility of the neutral atom and Δ is the difference of constants of the dispersive ion-atom interaction when the momentum projection on the internuclear direction is 0 and 1. Ion and atom temperatures were considered to be equal. As a result it was calculated that for the $4p^2P^0$ doublet $(9Q_i e^2 \alpha_a/2 + \Delta)/\hbar = 6\alpha_0 e^2 \langle r^2 \rangle_p / 5 = 365 \cdot 10^{-60}$ erg cm^6 and for the $4d^1\,{}^2P$ doublet - $2150 \cdot 10^{-60}$ erg cm^6.

By these means, the left hand side of equation (5.5) was experimentaly determined for different pressures and coordinates in the volume under study, and the right-hand side, which was λ - dependent, enabled us to estimate the λ parameters by means of the calculation of elements of the relaxation matrix and comparison with experimental measurements. Using

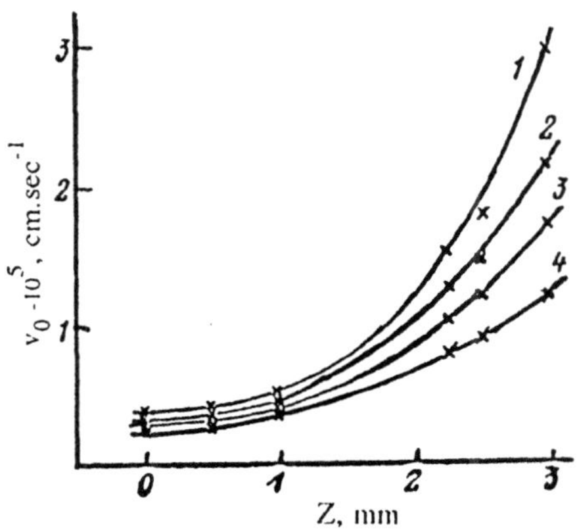

Figure 63. **Radial variation of the drift velocit y of argon ions at different pressures in the hollow cathode: P(Torr): (1) - 0.12, (2) - O.16, (3) - 0.2, (4) - 0.3**

the obtained λ parameters for two doublets, the drift velocity was finally obtained at every location and pressure in the gas under study (Fig. 63). The radial profile of the drift velocity is in accordance with the general concept of ion motion in the low pressure discharge plasma of a hollow cathode. In the center of the negative glow, the majority of the argon ions are formed as a result of direct impact ionization of the neutral component by electrons. Then they are accelerated in the radial direction and reach the surface of the cathode. Mass spectrometric studies confirm this picture.

Using the values of drift velocities mesured by this method and the mobility of ions in the gas, the internal electric field profile in the plasma was determined (Fig. 64) for pressures 0.1 - 0.3 Torr. The reliability of these data is confirmed by the comparison of the spectropolarimetric results with independent electric measurements of the potential drop on the hollow cathode. Linear extrapolation of the spectropolarimetric data to the dark space gives a magnitude of the electric field on the wall close to the independent direct electric measurements (Fig. 64). The principal criterion of reliability of the spectropolarimetric data obtained by the drift self-alignment technique is the agreement between the drift velocity data obtained for two different doublets in the boundary region of the negative glow.

The main advantage of the spectropolarimetric method is the possibility to use it for the measurement of low drift velocities. Choosing narrow spectral doublets, it is possible to measure very low drift velocities. The known spectroscopic techniques based on the Doppler shift measurement are limited by the requirement that the drift velocity should exceed the doppler broadening. In conclusion, this is a new remote sensing technique

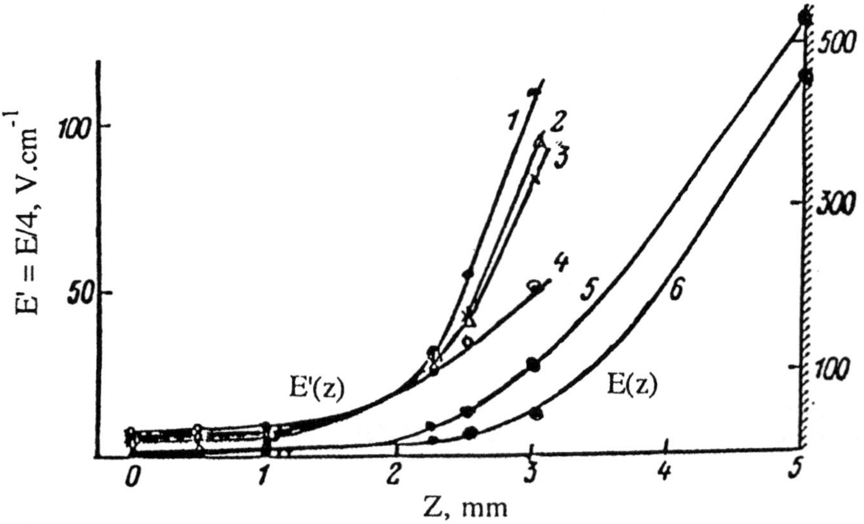

Figure 64. **Radial profile of electric field in the hollow cathode at various pressures P(Torr): (1) - 0.1, (2) - 0.16, (3) - 0.2, (4) - 0.3, (5) - 0.1, (6) - 0.3**

for the determination of the velocity of the charged particles in an ionized gas and of the distribution of the internal electric field that is based on the polarization of electron shells due to the anisotropy of the collisional relaxation.

CHAPTER 6

ATOMIC CONSTANTS DETERMINATION

The quantative application of the methods of spectropolarimetric sensing of ionized gases provides reliable information on the atomic characteristics of atomic states. In the present chapter we discuss applications of polarization spectroscopy of low temperature plasmas to the determination of lifetimes and depolarization collision cross-sections

6.1 Noble gases

Since it can be applied to the derivation of atomic constants, the polarization spectroscopy of gas discharges based on the Hanle technique extends beyond the optical spectropolarimetric sensing of the local physical conditions in laboratory or astrophysical plasma. The physical information incorporated in the magnetic depolarization profile is much richer than the one present in the Stokes parameters. For the most typical experimental schemes, the magnetic depolarization signal shape is described by a Lorentzian profile with a width determined by the alignment relaxation constant of the upper level of the spectral transition Γ_2. Therefore, the study of the self-alignment which arises in a plasma represents, in itself, an independent method for determination of atomic constants. An attractive feature of this technique is the possibility to study a large number of atomic states and in particular the highly excited states for which experimental data are limited, if even known. However, for reliable spectropolarimetric diagnostics /62, 83/, reliable quantitive spectroscopic data about the characteristics of the collisional depolarization of atomic states and measured lifetimes are indispensable.

6.1.1 RELAXATION CONSTANTS

All known methods of experimental determination of the relaxation constants of excited atomic states may be divided into three groups: spectroscopic methods based on the measurement of the spectral characteristics of emission, temporal analysis of the population decay of an excited state $\rho^{(0)}(t)$ and methods using the coherent effects of the quantum states of atomic ensembles and connected to the polarization moments of the higher rank $\rho_q^{(\kappa)}$ /62/. Hook method, total absorption technique, and spectral line intensity method are included in the first group of methods. Delayed coincidence technique, beamfoil and phase shift techniques form the second group. Interference beats method, Hanle and level crossing techniques, belong to the third group. The majority of experimental data on the determination of atomic lifetimes of excited states have been obtained by means of methods belonging to the second and third groups, i.e. delayed coincidence technique, beam-foil technique, and the method of level crossing.

A large number of experimental methods of coherent spectroscopy belonging to the third group have been elaborated by now /22/. The general feature of all these methods was the use of an external source of coherent excitation of an ensemble (resonance optical excitation, laser light excitation, collisional excitation by means of accelerated particles). Therefore, all known methods of coherent optics are connected to the development of special methods of excitation of the ensemble under study and the detection of optical radiation, which are usually rather complicated and may be used only for a limited number of states. Compared to the conventional techniques, polarization spectroscopy of ionized gases, based on the self-alignment of atomic states, offers an advantage, in that no special selective coherent excitation of different states is needed.

For low pressure gas discharge plasmas, the alignment relaxation constant Γ_2 is determined by the radiative lifetime of the level studied and the processes of interaction within the ensemble of atoms, such as depolarization collisions and radiation self-absorption /18, 22/:

$$\Gamma_2 = \Gamma - \Gamma_t + N_a \langle v_a \rangle \sigma_d + \Gamma^{(e)}_{coll} \tag{6.1}$$

where $\Gamma = (\sum_i W_i)^{-1}$ is the reciprocal radiation lifetime of the studied level, W_i is the transition probability from J, the level under study, to the lower level i, N_a is the concentration of the perturbing particles, $\langle v_a \rangle$ is the mean relative velocity of colliding particles, σ_d is the depolarization collision cross-section, $\Gamma^{(e)}_{coll}$ describes the relaxation of alignment by collisions with charged particles, and Γ_t stands for the radiation self-absorption.

Equation (6.1) is the basic relationship for the use of polarization spectroscopy for determining atomic constants using laboratory gas discharges. Experimental determination of the depolarization collision cross-sections by the Hanle technique in the gas discharge is based on the study of the dependence of the width of the self-alignment signal profiles $H_{1/2}$ as a function of the discharge parameters: the gas pressure and the discharge current. If radiation self-absorption and cascade transitions are not effective for the low pressure range ($p < 10$ Torr) and only binary collisions are valid, the relaxation constant Γ_2 is a linear function of the gas pressure (Fig .31a). Making use of the measurement of the linear pressure broadening of the self-alignment signal profile $\Delta\Gamma_2/\Delta p$ and of the relation (6.1), it is possible to determine σ_d the cross-section of the collisional destruction of alignment:

$$\sigma_d = \frac{4.46 \cdot 10^{-23}\sqrt{T}}{(M_1^{-1} + M_2^{-1})^{1/2}} \frac{\Delta\Gamma_2}{\Delta p}$$

Here M_1 and M_2 are the atomic masses of the colliding partners, $\Delta\Gamma_2/\Delta p$ is expressed in MHz/Torr and σ_d in cm^2.

The radiative lifetimes of excited states may be determined by extrapolating the widths of the self-alignment signals to zero pressure. For the majority of the excited states of noble gases, this extrapolation was linear reflecting the binary alignment destroying collisions with the ground state particles. Special studies showed that the nonlinear variation of Γ_2, within the low pressure range for different groups of levels, accounts for additional physical mechanisms such as resonance radiation diffusion within the plasma and cascade effects.

The collisional depolarization of states is the subject of a number of theoretical works, but there exists only a limited number of experimental measurements of the collisional depolarization cross sections. They are valid for noble gases and especially for the deep 2p states. Reliable experimental spectroscopic information on the collisional characteristics of the highly excited states are not known at all. It is the same for the lifetime determination. Many measurements of lifetimes have been carried out for the deep 2p states, but no reliable lifetime data are known for the highly excited states.

6.1.2 LIFETIMES AND ALIGNMENT-DESTROYING COLLISION CROSS-SECTIONS

Helium /47/. The polarization spectroscopy technique was applied to the positive column of an helium discharge maintained in a discharge tube of 6 mm bore, at pressure 0.1-0.5 Torr, with a 50-100 mA discharge current. Depolarization collision cross-sections and atomic lifetimes for some states presented in Table 6.1 have been measured.

TABLE 6.1. Lifetimes and depolarization collision cross section for some HeI states

	$3\,^1D_2$	$4\,^1D_2$	$5\,^1D_2$	$3\,^1P_1$	$4\,^1P_1$
τ, nsec	17 (2)	33 (7)	80 (40)	1,7	4,0 (4)
σ_d, Å^2	220 (30)	470 (60)	850 (130)	240 (20)	

Helium is the best example for testing new experimental techniques of measurement, because, from different measurements by reliable experimental methods and theoretical calculations, the spectroscopic characteristics of helium are known with a high degree of accuracy. Comparison of the $3, 4^1D_2$ level lifetimes derived from spectropolarimetric measurements showed agreement with the most reliable independent measurements. The optical depth of resonance radiation in the 0.1 - 0.5 Torr pressure range was rather high, so to obtain the radiative lifetimes for $3,4^1P_1$ levels from the apparent experimental lifetime τ_t, the theoretical relationship of these quantities /22/ was used :

$$\tau = \tau_t \left(1 - \frac{0.7W_2}{W_2 + W_n} \begin{Bmatrix} 1 & 1 & 2 \\ J & J & J_0 \end{Bmatrix}^2 \begin{Bmatrix} 1 & 1 & 0 \\ J & J & J_0 \end{Bmatrix}^{-2}\right), \qquad (6.2)$$

where $W_{r,n}$ are the transition probabilities for the resonance transition to the ground state of atom $J_0 - J$ and the $J_1 - J$ transition studied.

TABLE 6.2. Lifetimes and depolarization collision cross-sections for some NeI states

	$2p_2$	$2p_5$	$2p_6$	$2p_9$
σ_d, Å^2	16,4 (2,5)	50 (1)	17,3 (1,6)	92 (10)

	$2s_4$	$2s_5$	$2d'_1$	$3d_3$	$3d'_4$	$3d_4$
σ_d, Å^2	157 (30)	167 (30)	107 (20)	132 (20)	94 (28)	123 (30)
τ, nsec	8 (2)	66 (10)	22 (1)	25 (2)	21 (2)	26 (2)

	$3d_5$	$4d_2$	$3s_2$	$5s_4$	$4s'_1$
σ_d, Å^2	120 (90)	470 (40)	60 (30)	690 (160)	660 (60)
τ, nsec	13,1 (1,3)	21,5 (2,0)	23 (2)	57 (10)	25,8 (2,0)

Neon /53, 56/. Many experimental measurements of the depolarization collision cross-sections for the deep neon states are known. Some of these experiments utilized the Hanle technique under excitation by a He - Ne laser, which enables one to measure atomic constants of the generating levels /62, 22/. A number of experimental measurements of the deep 2p neon states have been realized by the self-alignment technique in a discharge plasma

/64, 62/, but the depolarization collision cross-sections for the highly excited neon states have been obtained only by the polarization spectroscopy of gas discharge /53, 56/. The use of the discharge self-alignment technique for the determination of the lifetimes of the deep $2p^5 3p$ levels was a more complicated problem. At a pressure above 0.5 Torr, the alignment relaxation constant is affected by radiation self-absorption in 1s-2p transitions. Accounting for this effect may be done under simplifyed assumptions and by using additional spectroscopic information. In view of that the self-alignment in the discharge plasma by the electron impact is used for the determination of the lifetimes of the highly excited states only when no distortions by the radiation self-absorption takes place. No experimental data for these levels were available before (see Table 6.2).

Argon /54, 56/. Determination of the depolarization collision cross sections of argon states by means of the polarization spectroscopy of the positive column of a gas discharge (0.2-3Torr, 20-100 mA, T_a = 340K) have been implemented and the results are presented in Table 6.3.

TABLE 6.3. Lifetimes and depolarization collision cross-sections for some ArI states

	$2p_2$	$2p_3$	$2p_4$	$2p_6$	$2p_7$
σ_d, Å^2	114 (10)	170 (20)	130 (10)	210 (20)	150 (20)

	$2p_8$	$2p_9$	$3p_{10}$	$3p_9$	$3p_8$
σ_d, Å^2	300 (10)	160 (20)	160 (90)	750 (90)	630 (30)
τ, nsec	–	–	150 (20)	185 (30)	135 (5)

	$3p_7$	$3p_6$	$4d_5$	$5d_5$	$6d_5$
σ_d, Å^2	1,07 (9) × 10^3	750 (300)	820 (400)	2,04 (0,2) × 10^3	3,9 (2) × 10^3
τ, nsec	130 (20)	140 (40)	–	–	–

Krypton /54, 56/. Table 6.4 gives the depolarization collision cross-sections for some krypton states.

TABLE 6.4. Lifetimes and depolarization collision cross sections for some KrI states

	$2p_2$	$2p_3$	$2p_6$	$2p_7$
σ_d, Å^2	690 (60)	160 (60)	940 (60)	240 (20)
τ, nsec	30 (4)	27 (1)	39 (3)	35 (1)

	$2p_9$	$3p_8$	$3p_7$	$3p_6$
σ_d, Å^2	170 (20)	1,0 (1).10^3	2,1 (2).10^3	880 (160)
τ, nsec	24 (2)	142 (7)	135 (5)	120 (20)

	$3p_2$	$4d_4$	$4d_3$
σ_d, Å^2	1,4 (1).10^3	1,7 (3).10^3	690 (60)
τ, nsec	132 (7)	210 (40)	125 (8)

	$5d'_4$	$6d'_4$	$3s_5$
σ_d, Å^2	3,2 (3).10^3	3,8 (3).10^3	1,3 (2).10^3
τ, nsec	220 (20)	180 (20)	63 (4)

Xenon /52, 56/. Depolarization cross sections and lifetimes of some xenon states measured by the spectropolarimetric technique are given in Table 6.5.

TABLE 6.5. Lifetimes and depolarization collision cross sections for some XeI states

	$2p_3$	$3p_8$	$3p_6$
σ_d, Å^2	500 (300)	1,8 (3).10^3	2,3 (3).10^3
τ, nsec	37,4 (8)	190 (30)	200 (20)
	$5d'_1$	$5d'_4$	$6d'_4$
σ_d, Å^2	2,8 (5).10^3	2,5 (3).10^3	3,5 (6).10^3
τ, nsec	200 (30)	125 (20)	180 (25)

6.1.3 CORRECTION FOR SELF-ABSORPTION

Study of systematic errors in the atomic constant determinations showed that errors for the highly excited states are determined mainly by the experimental signal to noise ratio and the statistical scatter of data. The linear variation of width of the signal profile and gas pressure is determined only by the depolarization collisions, and linear extrapolation to zero pressure gives the correct atomic constant values. Experimental estimates of absorption coefficients for the spectral lines corresponding to the transitions from this group of levels to the lower levels confirm this conclusion. The characteristic $\chi_0 d$ parameter is lower than 10^{-3}, which means that Γ_t is negligible compared to the real error bars of experiment.

For the deep 2p states of noble gases which are connected by strong optical transitions to 1s states, the situation is the opposite. For the pressure range beginning from 0.1 to several Torr, the characteristic $\chi_0 d$ parameter lies in the range 0.1 - 10. In this case it is impossible to neglect the correction for radiation self-absorption and a simple linear extrapolation of the widths of the self-alignment signals does not give correct values for radiative lifetimes.

The role of radiation self-absorption in the discharge plasma in the relaxation of the self-alignment was experimentally estimated in a spherical high frequency discharge in argon. The expression of Γ_t for a Doppler broadened line may be written in the form /18, 22/:

$$\Gamma_t = \sum_i W_i a_i (1 - \Pi_i),$$

where

$$\Pi_i = S(\chi_{0i} d) = \frac{1}{\sqrt{\pi}} \int_{-\infty}^{\infty} e^{-\nu^2} \exp\left(-\chi_{0i} d e^{-\nu^2}\right) d\nu.$$

Here χ_{0i} is the absorption coefficient in the center of a spectral line, d is the characteristic dimension of the discharge vessel,

$$\chi_{0i} = \frac{(2J+1)\lambda_i^3 N_i W_i}{(2J_i+1)8\pi}\left(\frac{M}{2\pi kT_a}\right)^{1/2};$$

λ_i is the wavelength of the radiation corresponding to the transition $J_1 - J$, N_1 is the concentration of the atoms in i state, and k is the Boltzmann constant and

$$a_i = \tfrac{7}{100}[3Y(Y-1)+8J(J+1)]^2/[(2J-1)2J(2J+2)(2J+3)]$$

$$Y = (J - J_i)(J + J_i + 1) + 2$$

In order to measure Γ_2, special additional experiments were performed using the spectral line absorption technique for the 1s-2p transition array. It was experimentally estimated that the error for 794.8, 706.7, 763.5, 801.4, 811.5 nm spectral lines caused by radiation self-absorption in the plasma exceeds 25 % and sometimes even more. Therefore, for the determination of atomic lifetimes, the widths of self-alignment signals were, in this case, measured simultaneously with the probabilities of radiation absorption in 1s-2p transitions. The correction for the self-absorption of radiation Γ_t was calculated and after that the quantity $\Gamma_2 - \Gamma_t$ was linearly extrapolated to zero pressure. The lifetimes of 2p states of argon measured by this complex technique are presented in the Table 6.6. These values are in good agreement with known lifetime data.

TABLE 6.6. Lifetimes of 2p argon states, measured by the magnetic spectropolarimetric technique in a high frequency discharge, taking into account radiation self-absorption in the plasma

	$2p_3$	$2p_4$	$2p_6$	$2p_8$	$2p_9$
τ, nsec	29 (3)	24 (1)	22 (1)	27 (1)	24 (1)

6.2 Electron beam excitation

In order to obtain reliable data for the lifetimes of the deep 2p and 3p states of noble gases by the spectropolarimetric technique, excitation was accomplished by a beam of low energy electrons. The attractive feature of this technique is the possibility to account for radiation self-absorption and cascade effects.

6.2.1 ACCOUNTING FOR THE BEAM MAGNETIC FIELD BENDING

The scheme of the spectropolarimetric experiment for electron beam excitation is presented in Fig. 65a. Atoms were excited in an electron gun consisting of an oxyde indirectly-heated cathode, three anodes and a collector. The mean electron current density in the beam was 9.5 mA/cm^2 and the radius of the cross-section of the beam was 1mm. Atoms of the noble gases were excited in the collision chamber between the third grid and the collector. The length of the collision chamber was $d_0 = 11.5$ mm. A linear polarizer was mounted before the entrance slit of the monochromator on which all radiation from the collision chamber was focused. The electron gun was placed inside Helmholz coils with an inner diameter of 10 cm and a constant 45 0e/A. The direction of the magnetic field was perpendicular to the electron beam and to the direction of observation. The maximal magnetic field strength was 10 0e, so that bending of the electron path took place as a result of the influence of the magnetic field on a moving electron. The electron beam was bended in the plane perpendicular to the entrance slit plane, not changing the position with respect to the entrance slit. The spectral line was projected to the photocathode of the photomultiplier working in the photon-counting regime with subsequent storage of information in the memory of the multichannel analyser. The switching of the channels of the multichannel analyser, operating in the slow temporal measurements regime, was synchronized with the stepwise variation of the magnetic field in the Helmholz coils.

For the description of the Hanle effect under the excitation of a beam of low energy electrons, let us use for the first stage the classical model of the linear oscillator, corresponding to a given spectral transition. The direction of oscillations was the same as the velocity vector of exciting electrons. The intensity of an ensemble of such oscillators in a magnetic field orthogonal to the oscillation direction is given by the formula:

$$I(\vec{e}) = I_0 \int_0^\infty dt\, e^{-\Gamma t} \cos^2 (\Omega t + \varphi),$$

where Γ is the damping constant of the oscillator, $\Omega = \gamma H$ is the Larmor frequency, γ is the gyromagnetic ratio, φ is the angle between the alignment axis at the moment of the excitation and the direction of the polarization vector $\vec{e}$.

This expression gives the standard Hanle signal shape which for $\varphi = 0, \pi/2$ corresponds to the Lorentzian dependence. The specific feature of the Hanle effect under electron impact excitation is the dependence of the angle φ on the magnetic field strength, because it is determined by the velocity vector of the exciting electron. It is clear that for every differentially small

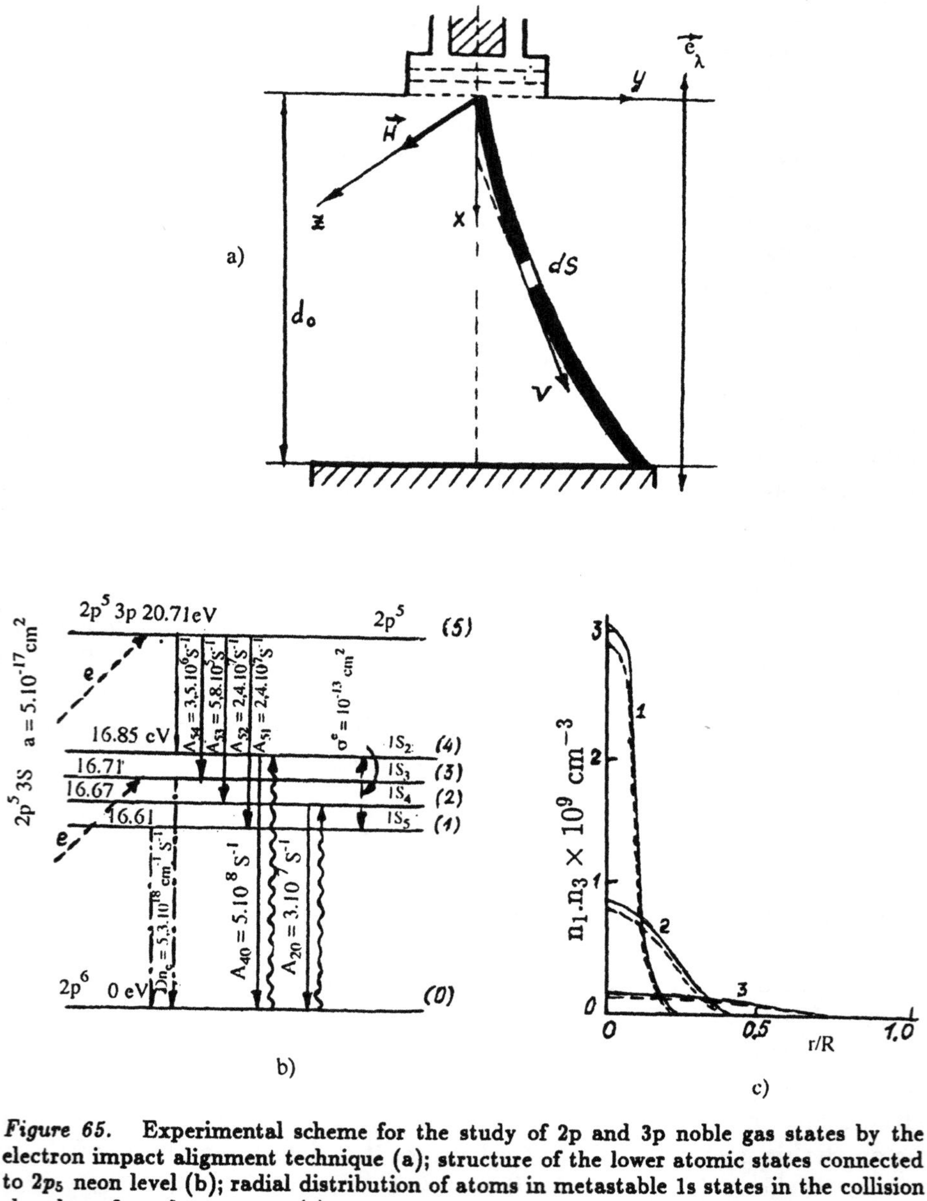

Figure 65. **Experimental scheme for the study of 2p and 3p noble gas states by the electron impact alignment technique (a); structure of the lower atomic states connected to $2p_5$ neon level (b); radial distribution of atoms in metastable 1s states in the collision chamber of an electron gun (c)**

part of the beam dl, the angle φ is 1-dependent: $\varphi = \varphi(l)$ (Fig. 65a). The $\varphi(l)$ dependence is easily calculated if the trajectory of the electron motion in the external magnetic field is known. The trajectory of an electron in the orthogonal homogeneous magnetic field is a circumference with the Larmor radius R_H. The divergence of the beam in the magnetic field was neglected because the electron beam was highly monokinetic. Neglecting also the deflection of the beam between the second and the third grids, we may write $\varphi(l) = 1/R_H$, where

$$R_H = \frac{c}{H}\sqrt{\frac{2m\varepsilon}{e^2}}$$

and finally:

$$I(\vec{e}) = I_0 \int_0^{L(H)} dl \int_0^{\infty} dt\, e^{-\Gamma t} \cos^2\left[\Omega t + l/R_H\right] \qquad (6.3)$$

The integration in (6.3) along the trajectory with the upper limit of the total length of the Larmor trajectory $L_H = R_H \arcsin d_0/R_H$ also accounts for the increase of the number of excited atoms in the collision chamber of the electron gun:

$$I(\vec{e}) = \frac{I_0}{2\Gamma}\left[R_H \arcsin d_0/R_H + d_0\sqrt{1 - \frac{d_0^2}{R_H^2}\frac{1}{1+x^2}} - \frac{d_0^2}{R_H}\frac{x}{1+x^2}\right], \qquad (6.4)$$

where $x = 2\Omega/\Gamma$.

For the polarization moment formalism, the intensity of the light with polarization $\vec{e}$ is determined by the expression (1.24). For the given experimentral geometry (Fig. 65a), the observation tensor $\Phi_q^{(\kappa)}(\vec{e})$ consists of the following components:

$$\Phi_0^{(0)} = -1/\sqrt{3} \quad \Phi_0^{(1)} = 0 \qquad \Phi_0^{(2)} = -1/\sqrt{30}$$
$$\Phi_1^{(2)} = \Phi_{-1}^{(2)} = 0 \qquad \Phi_2^{(2)} = \Phi_{-2}^{(2)} = -1/2\sqrt{3}.$$

The polarization moments characterizing the atomic subensemble corresponding on the differential element, dl, of the electron trajectory are expressed by:

$$d\rho_q^{(\kappa)} = \sum_{q'} D_{q'0}^{(\kappa)}[0, \pi/2, \varphi(l)]\, d\rho_0^{(\kappa)}$$

where $D_{q'0}^{(\kappa)}[0, \pi/2, \varphi(l)]$ is the Wigner matrix of rotation.

Accounting for the radiation and collisional relaxation of polarization moments and the influence of the magnetic field we have:

$$dI(\vec{e}) = a' + b'\frac{\Gamma_2 \cos 2\varphi(l) - 2\Omega \sin 2\varphi(l)}{\Gamma_2^2 + 4\Omega^2}$$

where a' and b' are numerical coefficients.

Integrating along the arc corresponding to the electron trajectory in the collision chamber magnetic field, we obtain for the linear polarization of the spectral line of wavelength λ the same expression as (6,4):

$$I(\vec{e}) = a'' \arcsin\left(\frac{d_0}{R_H}\right) + b'' \left(d_0\sqrt{1 - \frac{d_0^2}{R_H^2}\frac{1}{1+x^2}} - \frac{d_0^2}{R_H}\frac{x}{1+x^2}\right) \quad (6.5)$$

where a'' and b'' are numerical constants.

These formulae have been applied to the calculation of Hanle signal profiles resulting from excitation in the collision chamber of an electron gun ($d_0 = 11.5$ mm). The determination of relaxation constants was carried out by comparing of the theoretical curve (6.4) with the experimental profile measured for different gas pressures, beam current densities and electron energies. Radiative lifetimes were obtained by extrapolation of relaxation constants to zero pressure.

6.3.2 RADIATION SELF-ABSORPTION

The characteristic feature of radiation self-absorption in the collision chamber of the electron gun is that the distribution of absorbing atoms is not uniform. The maximal density of 1s atoms is within the electron beam (Fig. 65)) and it is decreasing while moving towards the wall of the collision chamber because of the metastable diffusion in the own gas atmosphere

.

Estimates of Γ_t have been made within the framework of the simplified approach for the calculation of the optical depth of the collision chamber for the transition under study, changing (2.15) to:

$$\chi_{0i} = \frac{(2J+1)\lambda_i^3 W_i R^{-1}}{2\pi(2J_i+1)}\sqrt{\frac{M}{2\pi k T_a}}\int_0^R N_i(r)dr.$$

The calculation of Γ_t for 2p states of noble gases on the first stage includes the calculation of the radial distribution of the absorbing atoms in the collision chamber /84/. A typical level scheme is given in Fig. 65b for the $2p_5$ neon level, where the effective constants describing different processes (radiative decay, metastable diffusion, collisional mixing within 1s configuration) are presented. Supposing that the excited atom concentration is much less than that for ground state atoms, we have the following system of population balance equations:

$$\begin{aligned}
\frac{\partial N_1}{\partial t} = {} & F_1 + D\Delta N_1 + (k^a_{12} + k^e_{12})N_2 + (k^a_{13} + k^e_{13})N_3 + \\
& (k^a_{14} + k^e_{14})N_4 - (k^a_{21} + k^e_{21} + k^a_{31} + k^e_{31} + k^a_{41} + k^e_{41})N_1 + \\
& W_{51}N_5 - W_{51}N_1 \int dr' N_5(r')\chi_{15}(rr') \\
\frac{\partial N_2}{\partial t} = {} & F_2 + D_1\Delta N_2 + (k^a_{21} + k^e_{21})N_1 + (k^a_{23} + k^e_{23})N_3 + (k^a_{24} + k^e_{24})N_4 - \\
& (k^a_{12} + k^e_{12} + k^a_{32} + k^e_{32} + k^a_{42} + k^e_{42})N_2 + W_{52}N_5 \\
& -W_{52}N_2 \int dr' N_5(r')\chi_{25}(rr') - W_{20}N_2 + \\
& W_{20}N_0 \int dr' N_2(r')\chi_{02}(rr') \\
\frac{\partial N_3}{\partial t} = {} & F_3 + D\Delta N_3 + (k^a_{31} + k^e_{31})N_1 + (k^a_{23} + k^e_{23})N_2 + \\
& (k^a_{34} + k^e_{34})N_4 - (k^a_{13} + k^e_{13} + k^a_{23} + k^e_{23} + k^a_{43} + k^e_{43})N_3 + \\
& W_{51}N_5 - W_{53}N_3 \int dr' N_5(r')\chi_{35}(rr') \\
\frac{\partial N_4}{\partial t} = {} & F_4 + D_1\Delta N_4 + (k^a_{41} + k^e_{41})N_1 + (k^a_{42} + k^e_{42})N_2 + (k^a_{43} + k^e_{43})N_3 - \\
& (k^a_{14} + k^e_{14} + k^a_{24} + k^e_{24} + k^a_{34} + k^e_{34})N_4 + W_{54}N_5 \\
& -W_{54}N_4 \int dr' N_5(r')\chi_{25}(rr') \\
& W_{40}N_0 \int dr' N_4(r')\chi_{04}(rr') \\
\frac{\partial N_5}{\partial t} = {} & F_5 + D_2\Delta N_5 - (W_{51} + W_{52} + W_{53} + W_{54})N_5 + \\
& W_{51} \int dr' N_1(r')N_5(r')\chi_{15}(rr') + \\
& W_{52} \int dr' N_2(r')N_5(r')\chi_{25}(rr') + \\
& W_{53} \int dr' N_3(r')N_5(r')\chi_{35}(rr') + \\
& W_{54} \int dr' N_4(r')N_5(r')\chi_{45}(rr')
\end{aligned} \tag{6.7}$$

where $F_J = NN_e v\sigma_J f(r)$ is the rate for the electron impact excitation, $f(r) = 1(r \leq R^*)$, $f(r) = i_p/i_e\,(R^* < r < R)$, i_e, i_p are electron densities within the beam of radius R^* and in the rest volume of the collision chamber. The second term in the balance equations describes diffusion, D, D_1 , and D_2 are diffusion coefficients of the metastable and radiating states of 1s and 2p configurations.

The third term stands for the intramultiplet collisional mixing induced by collisions with neutral atoms, $k^a_{iJ} = N\langle v_a\rangle\sigma^a_{iJ}$, and electrons, $k^e_{iJ} = N_e v\sigma^e_{iJ} f(r)$. The integral term represents self-absorption of radiation on $i-j$ transitions. It is possible to write:

$$N_J(r')\chi_{iJ}(rr') = \frac{\chi_{0iJ}(r')}{4\pi \mid \vec{r} - \vec{r'} \mid^2} \int d\nu\, \varepsilon_{iJ}(\nu)\tilde{\chi}_{iJ}(\nu) e^{-\tilde{\chi}_{iJ}(\nu)\int_r^{r'} dr''\, \chi_{0iJ}(r'')}$$

where $\chi_{0iJ}(r)$ is a linear function of $N_J(r')$; the Doppler profile is determined by (6.6), where $\varepsilon_{iJ}(\nu)$ and $\tilde{\chi}_{iJ}(\nu)$ are emission and absorption line profiles coefficients.

The boundary condition for (6.7) system is $N_J \mid_{r=R} = 0$, which means that collisions with the wall destroy all the excited atoms. The solution of this system was used to estimate the total quantity of atoms in a given cross-section of the excitation chamber:

$$\bar{N}_J = 2\pi \int_0^R dr\, r N_J(r)$$

Results of such calculations are presented in Fig. 65 and in Table 6.7.

TABLE 6.7. Calculated contribution to alignment relaxation constant (in MHz) caused by the radiation self-absorption in the collision chamber of the electron gun for some states of noble gases at different pressures; (I) - with account for $np^6 - np^5(n+1)s$ radiation self-absorption, (II) - with no account of it)

p, Tor		Kr $3p_2$	Kr $3p_8$	Xe $2p_2$	Xe $2p_4$	He $4\,^1D_2$	He $5\,^1D_2$
10^{-2}	I	1,2	6,83	$9{,}86.10^5$	$5{,}15.10^5$	$5{,}75.10^1$	2,5
	II	$9{,}14.10^{-1}$	6,24	$9{,}81.10^5$	$5{,}10.10^5$		
5.10^{-2}	I	$6{,}90.10^1$	$2{,}42.10^2$	$8{,}82.10^6$	$5{,}41.10^6$	$1{,}05.10^4$	$4{,}62.10^2$
	II	$2{,}04.10^1$	$1{,}39.10^2$	$8{,}58.10^6$	$5{,}18.10^6$		
10^{-1}	I	$2{,}84.10^2$	$9{,}56.10^2$	$1{,}06.10^7$	$7{,}04.10^6$	$4{,}37.10^4$	$1{,}94.10^3$
	II	$7{,}18.10^1$	$4{,}87.10^2$	$9{,}75.10^6$	$6{,}32.10^6$		

For all 2p-neon levels studied, and for $3p_2, 3p_8$ krypton and $4^1D_2, 5^1D_2$ helium levels, Γ_t was small. For example, for $2p_5$ neon level at 0.1Torr, $\Gamma_t = 0.006\Gamma_0$. The role of radiation self-absorption was very pronounced for 2p xenon levels because of the high oscillator strengths for 1s - 2p transitions and of the considerable cross-sections of 1s levels by electron impact excitation. So for a pressure of 0.1 Torr, $\Gamma_t(2p_4Xe)$ was 28% Γ_0. For 4^1P_1 helium level, the estimate for the case of strong radiation self-absorption was made with the help of expression (6.2).

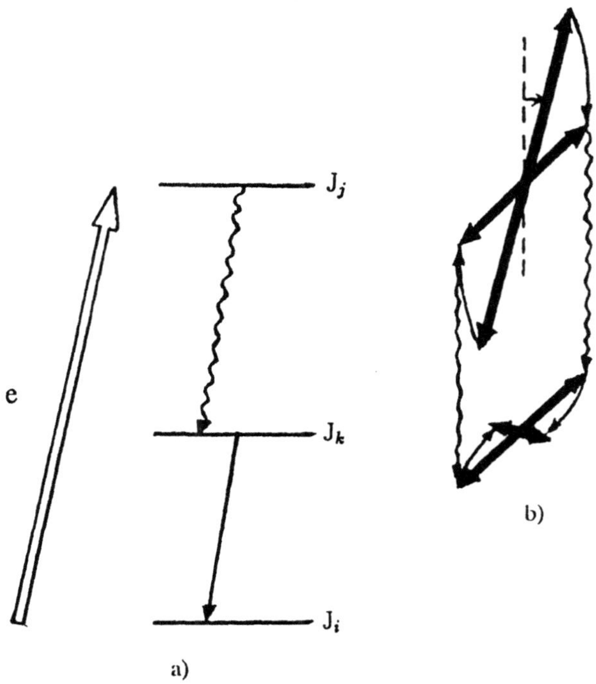

Figure 66. **Schematic illustration for the calculation of the cascade alignment signal profile under electron impact excitation for three-levels (a); precession of atomic oscillators in an external magnetic field (b)**

6.2.3 ALIGNMENT CASCADING

It is known that, under nonselective electron impact excitation of the deep energy states of atoms and ions, the cascade transitions are the main error source for lifetime determinations. The influence of alignment cascading in radiative lifetime determinations under nonselective electron impact excitation may be estimated by analysing the cascade profile of Hanle signal. For the simple three-level scheme, it is possible to use the model of the classical oscillator, precessing in an external magnetic field. Let the moment $t_j = 0$ correspond to the excitation of the high-lying level j, the transition to the studied level k takes place at the instant t, and t_k be the time interval during which the oscillator, corresponding to i - k transition is radiating (Fig. 66). Assuming that relaxation constants of j and k oscillators are different, ($\Gamma_j \neq \Gamma_k$) and that the precession frequencies are also different, we may write the expression for the rotation angle of the oscillator axis in the magnetic field: $\Omega_j t_j + \Omega_k t_k + \varphi$, where Ω_j and Ω_k are Larmor frequencies of precession of the atomic oscillators, corresponding to j and k atomic states and φ is the initial phase of oscillations at $t_j = 0$. Accounting for magnetic field deflection of the electron trajectory, the dependence of linear polarization of the spectral line takes the form:

$$I(\vec{e}) = I_0 \int_0^{L_H} dl \int_0^{\infty} dt_j \, e^{-\Gamma_j t_j} \int_0^{\infty} dt_k \, e^{-\Gamma_k t_k} \cos^2 \left[\Omega_j t_j + \Omega_k t_k + \frac{l}{R_H} \right]$$

where l is the arc length of the circumference with Larmor radius

$$R_H = c\sqrt{2m\varepsilon / eH^{-1}}$$

describing the electron trajectory in the collision chamber with the energy ε, and

$$L_H = R_H \arcsin d_o / R_H$$

is the total length of the trajectory of an electron in the collision chamber. Integrating for $d_o \ll R$ we obtain:

$$\begin{aligned} I(\vec{e}) \quad = \quad & \tfrac{I_0}{2\Gamma_k\Gamma_j} R_H \arcsin(d_o : R_H) + \tfrac{I_0}{2\Gamma_k\Gamma_j} d_o \sqrt{1 - \tfrac{d_o^2}{R_H^2}} (x_j x_k - 1)/((1 + x_j^2)(1 \\ & - \tfrac{I_0}{2\Gamma_k\Gamma_j}((d_o^2(x_j - x_k))/(R_H(1 + x_j^2)(1 + x_k^2)), \end{aligned}$$

where $x_{j,k} = 2\Omega_{j,k}/\Gamma_{j,k}$.

Generalization of this expression for the case of many cascade transitions is easily done using the formalism of polarization moments. Within the frame of reference (Fig. 65a), the expression for polarization moments of the subensemble of atoms in the state k which corresponds to the arc length dl will take the form:

$$d\rho_{qk}^{(\kappa)} = \sum_j (-1)^{J_j + J_k + \kappa + 1} W_{jk} (2J_j + 1) \left\{ \begin{array}{ccc} J_j & J_j & \kappa \\ J_k & J_k & 1 \end{array} \right\}$$

$$\frac{\sum_{q'} D_{q'0}^{(\kappa)} \left(0, \frac{\pi}{2}, \frac{l}{R_H}\right) \, {}^{\vec{n}}\rho_0^{(\kappa)j}}{(\Gamma_j - 2i\Omega_j)(\Gamma_k - 2i\Omega_k)}.$$

Here $J_{j,k}$ are the angular momenta of electron states j, k; $W_{j,k}$ is the transition probability k - j, ${}^{\vec{n}}\rho_0^{(\kappa)j}$ is the polarization moment of j state in the frame of reference with the quantization axis oriented along the velocity vector, and $D_{qq'}^{\kappa}\left(0, \frac{\pi}{2}, \frac{l}{R_H}\right)$ is the Wigner matrix accounting for the transfer from the collision frame of reference to the laboratory frame of reference.

The total intensity of light with linear polarization $\vec{e}$ takes the form:

$$I(\vec{e}) = B^* \int_0^L dl \sum_{\kappa,q} (2\kappa + 1) \left\{ \begin{array}{ccc} 1 & 1 & \kappa \\ J_k & J_k & J \end{array} \right\} (-1)^q d\rho_{qk}^{(\kappa)} \Phi_{-q}^{(\kappa)}(\vec{e})$$

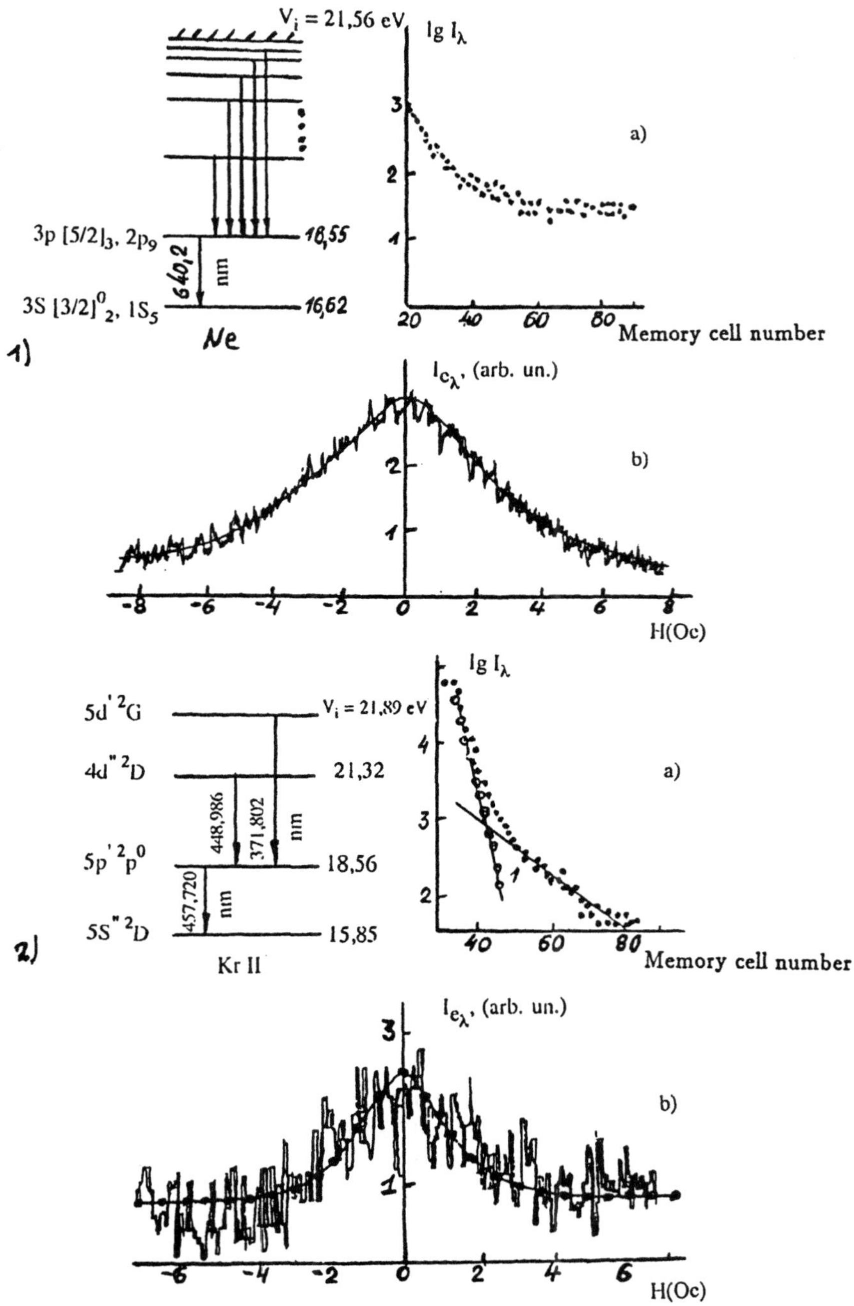

Figure 67. **Alignment signal profiles under electron impact excitation (1b, 2b), population decay curves (1a, 2a) for large (1 - Ne) and small (2 - Ionized Krypton) numbers of cascade channels**

where $\Phi^{(\kappa)}_{-q}(\vec{e})$ is the observation tensor, expressed by the formulae (1.24) and B^* is a constant.

Finally for $I(\vec{e})$ we have:

$$I(\vec{e}) = I^{[1]}(\vec{e}) + I^{[2]}(\vec{e}) \tag{6.8}$$

where $I^{[1]}(\vec{e})$ is the Hanle signal shape resulting from the direct excitation of the level (6.4) and

$$\begin{aligned} I^{[2]}(\vec{e}) \quad = \quad & B^* \textstyle\sum_j (2J_j + 1) W_{jk} \\ & \left[a''_j R_H \arcsin \frac{d_o}{R_H} + (-1)^{J_j - J_k + 3} b'' \left\{ \begin{matrix} J_j & J_j & 2 \\ J_k & J_k & 1 \end{matrix} \right\} \mathcal{L}_{\rm j} \right] . \end{aligned}$$

Here a'', b'' describe the excitation and alignment of the state j by electron impact and $\mathcal{L}_{\rm j}$ is given by:

$$\mathcal{L}_{\rm j} = {\rm d}_0 \sqrt{1 - \frac{{\rm d}_0^2}{{\rm R}_{\rm H}^2} \frac{{\rm x_k x_j} - 1}{(1 + {\rm x_j^2})(1 + {\rm x_k^2})}} - \frac{{\rm d}_0^2}{{\rm R}_{\rm H}} \frac{{\rm x_j + x_k}}{(1 + {\rm x_j^2})(1 + {\rm x_k^2})} .$$

The profile $I^{[2]}(\vec{e})$ represents the sum of contributions, reflecting every cascade channel of the state k. The sign of every contribution is determined by the momentum of the upper j and the studied state k. It means that, because of cascade compensation, for a large number of cascade transitions k - j, the total value of $I^{[2]}(\vec{e})$ may become much smaller with respect to $I^{[1]}(\vec{e})$ ($I^{[2]}(\vec{e}) \to 0$) .

Experimental proof for the multicascade compensation of alignment was obtained in lifetime measurements by the combination of the Hanle technique and of the delayed coincidence method, applied to the same object. This experiment was based on the fact that the population of the highly excited states under cascading may be transformed only into the population. So the decay curve measured by the delayed coincidence technique may be used to determine the scheme of the cascade population of the studied level as well as the share of the cascade population. For this purpose, the electron gun was operated in the pulse regime. The pulsed signal from the photomultiplier entered the time-amplitude converter with a time scale from 100 nsec to 10 μsec and then entered the multichannel pulse analyser. The electron gun was operated by the pulsed signal of duration of 10-200 nsec, which was the start pulse for the time-amplitude converter. The signal of the Hanle effect was detected under the continuous electron impact excitation. The linear polarizer was mounted in the observation channel and the magnetic field was applied to the collision chamber of the electron gun. It was periodically changed according to a stepwise saw-tooth law. Information was stored in the memory of the multi-channel analyser, the memory address of which was changed according to the variation of the magnetic field strength.

Experimental studies have been performed on 2p neon states and some states of the singly ionized krypton. Decay curves of $2p_5, 2p_6$ and $2p_9$ levels showed strong cascade population for the energy of exciting electrons considerably exceeding the threshold value. Special studies showed that the decay curve of $2p_9$ neon level includes 13 cascade transitions. The Hanle signal profile for the same neon states did not show any cascade components (Fig. 67) and the lifetimes which have been obtained using the theoretical expression (6.4) were in agreement with the experimental threshold data.

TABLE 6.8 Lifetimes of some atomic states of noble gases, measured by magnetic spectropolarimetric technique under electron beam excitation

Néon

	$2p_2$	$2p_4$	$2p_5$	$2p_6$	$2p_8$
τ, nsec	19,5 (1,3)	19,7 (6)	19,4 (8)	19,1 (1,2)	17,5 (1,0)

Argon

	$3p_3$	$3p_4$	$3p_6$	$3p_7$	$3p_9$
τ, nsec	130 (30)	150 (10)	155 (20)	150 (20)	160 (30)

Krypton

	$2p_2$	$3p_2$	$3p_6$	$3p_7$	$3p_8$
τ, nsec	28 (2)	120 (30)	120 (30)	126 (10)	130 (15)

Xénon

	$2p_2$	$2p_3$	$2p_4$	$3p_7$	$3p_9$
τ, nsec	27 (1)	36 (1)	34 (1)	200 (40)	160 (30)

This result may be explained only if multicascade compensation of alignment occurred. This idea was confirmed by observations on some ionic states of krypton, where the opposite cascade situation took place. Population decay curves under nonselective electron impact excitation were represented as a sum of one or two exponents. The numerical processing of these signals by the least square technique indicated that the cascade share did not exceed 0.5 - 1.5% of the total population of an electron energy level of 30 ev. Lifetimes of the states measured by this technique were in agreement with those measured under threshold excitation conditions, confirming the correctness of the cascade scheme in this case.

The Hanle signal shape also indicated a pronounced narrow cascade component. The population decay curve indicated that there were only one or two cascade transitions and that cascade compensation was impossible. Relaxation constants of the highly excited states which have been obtained from the resolution of the decay curves were used for calculating the cascade Hanle profile by means of the formula (6.8). A comparison of experimental and theoretical profiles of the cascade component is given in Fig. 67. Agreement between the theoretical curve and the experimental data confirm the role of cascading in the kinetic of population and alignment of an atomic state under the nonselective excitation, as well as the multicascade alignment compensation. Experimental data for lifetimes of some states of the noble gases are given in the Table 6.8.

The effect of the multicascade alignment compensation may be of importance for the determination of atomic constants under nonselective impact excitation of atomic states. In such cases, the polarization spectroscopy methods are not affected to the same degree by cascade distortions with respect to the intensity methods.

CHAPTER 7

POLARIZATION SPECTROSCOPY IN ASTROPHYSICS

The polarization of radiation from the outer part of the solar atmosphere has been under continuous study for many years /85, 86/. The polarization of radiation coming from the solar photosphere is used daily by Solar Observatories all around the world to measure the spatial distribution of the solar photospheric magnetic field of solar active regions. During solar eclipses and with devoted instruments, where the radiation from the solar disk is occulted in order to allow observations of the much weaker emission from the low corona, the polarization of the radiation from the corona and from external concentration of neutral gas (prominences) is observed and studied. Using the most simple and straightforward model of Thomson scattering of light from the photosphere by free electrons and the theory of resonant scattering by free atoms and ions, the observations of the continuum and line spectra of the corona provide information on physical conditions in the solar corona. More recently, linear line polarization generated by anisotropies in the collisional excitation conditions of atomic species has been observed during solar flares, providing additional information on particle beams.

7.1 Solar magnetic field measurement

Long ago, polarized optical emission of the outer parts of the solar atmosphere was detected. On the basis of developed quantum mechanical methods it soon became possible to build up quantitative polarization spectroscopic methods, considerably expanding the horizons of ground-based observational techniques. Measurements of the polarization of radiation from the solar atmosphere have long been used for magnetometry based on Zeeman and Hanle effects.

7.1.1 ZEEMAN AND HANLE EFFECTS

Zeeman Effect

Most solar magnetometers work by means of the Zeeman effect, i.e. the splitting of an emission line into elliptically polarized components in a magnetic field. The nature of the polarization varies from circular to linear depending on the orientation of the magnetic field relative to the line of sight direction. The Zeeman splitting of a line depends on the nature of the atomic transition and, except for very high fields, is proportional to the magnetic field intensity. If the splitting is less than the width of the line, the degree of polarization in different parts of the line contour is measured to determine the magnetic field from the Zeeman effect. Such measurements require observations with high spectral resolution. Zeeman magnetometry is used mainly to measure the magnetic field of the solar photosphere. Measurement of line circular polarization yields reliable values of the longitudinal component of the magnetic field which is parallel to the line of sight. The longitudinal component of the solar photospheric magnetic field can be measured down to a few Gauss. The transverse component can be derived from measurements of the Zeeman linear polarization. Reliable measurements of the transverse component are restricted to fields higher than about 100 Gauss in active regions. Since application of the Zeeman effect to the measurement of the vector magnetic field in the solar photosphere is described in detail in /5/ it is not further developed here.

Hanle Effect

Using atomic lines formed by resonance, transverse weak magnetic fields can be measured using the Hanle effect. The Hanle effect is the rotation of the polarization plane of scattered resonance radiation under the effect of a magnetic field (see section 1.1.1). Measurements by means of the Hanle effect do not require a high spectral resolution, since only the integrated contour of the spectral line is measured. It is thus possible to improve the spatial resolution of the method and to study fainter spectral lines. This method has one other important advantage in addition to its experimental simplicity. Measurements of two spectral lines make it possible to reliably determine the local magnetic field vector. However, it should be noted that the Hanle-effect method is applicable only in cases in which the conditions in the ensemble of atoms under study correspond to self-alignment. In case when self-absorption is induced by reabsorption of resonance optical radiation, it is possible on the basis of geometric considerations to describe the anisotropy of the optical radiation. Then, using additional information about the processes responsible for the formation of the spectral line, and about the characteristics of the medium of the object under study, the polarization moments of the upper level of the transition can be calculated. These moments determine the polarization characteristics of the radiation from the ensemble of particles in the presence of a magnetic field of given orientation and strength. Then the local magnetic field can be determined by comparing the measured and calculated values.

An important feature of a Hanle-effect magnetometer is its sensitivity to the strength of the magnetic field being measured. The sensitivity is greatest under the condition $1 \leq \omega\tau^{(2)} \leq 10$, i.e., when ω, the Larmor precession frequency of the atomic oscillator in the magnetic field, is approximately equal to the corresponding relaxation constant $1/\tau^{(2)}$, where $\tau^{(2)}$ is the relaxation time of self-alignment. Under this condition, the radiation may contain a noticeable magnetic field-induced rotation of the polarization plane, and the degree of polarization may be reduced. In the case $\omega\tau^{(2)} \gg 1$, the degree of polarization drops to a minimum value independent of the magnetic field. Conversely, if $\omega\tau^{(2)} \ll 1$, the magnetic field does not cause any significant change in the direction or degree of polarization of the radiation.

Spectral lines suitable for Hanle-method measurements were specially selected, allowing for the typical magnetic fields in the various parts of the solar atmosphere. In prominences the magnetic field is of the order of a few Gauss. Therefore, the lifetimes of the excited atoms used to measure the field must be of the order of 10^{-7} s. For the corona these fields are lower, generally $10^{-4} - 10^{-5}$ Gauss, so that the method can be implemented if the relaxation constant of the upper level of the transition of interest lies in the range $10 \leq 1/\tau^{(2)} \leq 5000$ s^{-1}. This requires the use of lines of low transition probability, such as forbidden lines.

7.1.2 SOLAR PROMINENCES MAGNETIC FIELD

The first indication of polarization of the spectral lines emitted from the outer part of the solar atmosphere, due to absorption of anisotropic resonance radiation from the inner parts of the atmosphere, was reported by Ohman /87/, who predicted polarization of the radiation in the spectral lines of hydrogen and ionized calcium in the radiation from high prominences. A qualitative picture of the polarization state of the radiation from a prominence can be drawn from a classical examination of the alignment process (see 1.1.1). In a prominence excited by photospheric light, in the absence of a magnetic field, the vertical axis OZ is a symmetry axis (Fig. 68). As a result, observation in the OY direction will reveal radiation which is partially polarized in the OX direction. The polarization of the radiation is described in classical terms as the radiation from a linear oscillator whose oscillation axis is parallel to OX, i.e. tangent to the solar limb. Due to an increase in the anisotropy of the optical excitation of regions far from the solar limb, the degree of polarization will increase with height.

Many subsequent experimental observations and studies of the linear polarization of the radiation from prominences were then carried out and the polarization of the radiation in the continuum, in the H_α line of hydrogen and in the D_2 line of helium, was measured. It was found that there is a systematic deviation of the polarization directions of these lines from the tangent to the solar limb. The deviation angles were found to lie in the range $11 - 30^0$ for H_α and $22 - 44^0$ for D_3 lines.

These experiments spurred some early theoretical work, in which the polarization of the radiation from an ensemble of atoms in a solar prominence

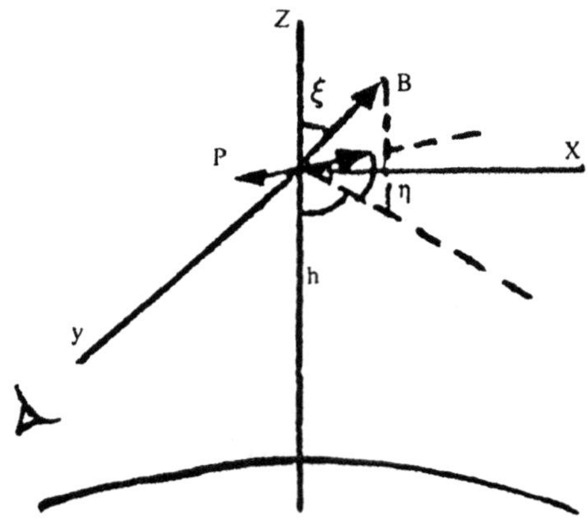

Figure 68. Coordinate system for describing self-alignment in the solar atmosphere

was studied in the classical model of an atomic oscillator, using a more realistic model of anisotropic photoexcitation. Including a limb-darkening effect, Zanstra /88/ calculated the degree of polarization of the hydrogen H_α and helium D_3 lines at various points in a prominence with a small optical density . The observed deviation from the tangential direction of the direction of the predominant oscillation of the electric vector in the scattered light was explained /89/, on the basis of the Hanle effect, as an effect of the local magnetic field at the observation point. The local magnetic field $B(\xi, \psi)$ (Fig. 68) leads to a precession of the oscillator. As a consequence, the polarization plane of the radiation rotates, and there is a decrease in the degree of polarization. The magnitude of the effect depends not only on the magnitude of the magnetic field but also on its spatial orientation, which is specified by the angles ξ and ψ. If the magnetic field is along the OZ axis, it does not change the polarization of the scattered light, since it is directed along the symmetry axis of the optical excitation of the ensemble of atoms. A magnetic field along the OX axis simply reduces the degree of polarization, leaving the direction of the predominant oscillation of the vector E in the scattered light wave unchanged. A longitudinal magnetic field along the OY axis both reduces the degree of polarization and rotates the polarization direction.

Many experimental observations of the polarization of the radiation from prominences in spectral lines were carried out in the 1960s and 1970s. The model of anisotropic resonant optical excitation was used to interpret the observations, with the goal of determining the local magnetic field. Hyder /90, 91/ studied the linear polarization of the radiation from a loop prominence in the bright upper parts of the loops. It was found that the degree of polarization of the radiation in the hydrogen H_α line was 0.8 %, and the deviation angle from the tangential direction was $20 - 25^0$. Under the assumption that the prominence had a small optical thickness, these experimental results led to magnetic fields of the order of 45-60 Gauss. Warwick and Hyder /92/ studied the polarization of the radiation in the helium D_3 line. They also carried out calculations for an optically thin collisionless prominence. The anisotropic photoexcitation of the lower level of the atomic transition of interest, ^3P, by radiation from the photosphere in the spectral line at 10 830 Å, corresponding to the helium atom transition $^3S - {}^3P$, and the effect of multiple scattering of photons in the medium of the prominence were also taken into account. Hyder reconstructed the magnetic field configuration and estimated the field on the basis of polarimetric data.

The polarization characteristics of the 4227 Å CaI line in the radiation from a quiet prominence were studied /93/. At an altitude of 10 000-15 000 km, the degree of polarization reached 4%, i.e. less than that calculated in the model of anisotropic photoexcitation. The deviation was attributed to two factors: an additional excitation of the state of interest by isotropic collisions with electrons and the effect of the local magnetic field, whose magnitude was estimated from the Hanle effect to be 25 Gauss.

The possibilities of exploiting the Hanle effect in radiation from quiet prominences for magnetometry were examined in detail in a series of studies, based on extensive observations, including more than 1000 measurements at Pic du Midi Observatory of the polarization of the radiation from 90 prominences of various types between December 1973 and May 1976. The results of the observations in the helium D_3 line (5876 Å) were summarized by Leroy *etal* /94/. This line has a rather high radiation intensity, and the results /95/ indicate that the primary mechanism for the excitation of the D_3 line in prominences is the absorption of resonance radiation from the photosphere. In this case, even for bright prominences, the approximation of an optically thin layer can be used, substantially simplifying the interpretation of the results of polarimetric measurements. Fig. 69 shows histograms of the distribution, in the scattered light of a quiet prominence, of the observed degree of polarization, and of the angle of rotation of the direction of predominant oscillation vector E with respect to the tangential direction. The most typical value of the degree of polarization is in the range 1 to 3.5% while the rotation angle is $\pm 15^o$; these quantities vary slightly over the surface of a prominence.

A detailed study was made of how the degree of polarization, the maximum polarization, the deviation angle, and the intensity of the spectral line depend on the altitude of the region of the quiet prominence under study with respect to the solar limb. Analysis of the results confirmed that the magnetic field affects the polarization through the Hanle effect, so that the results can be used to determine the magnetic field. In this series of studies, the local magnetic field B and its spatial orientation, specified by

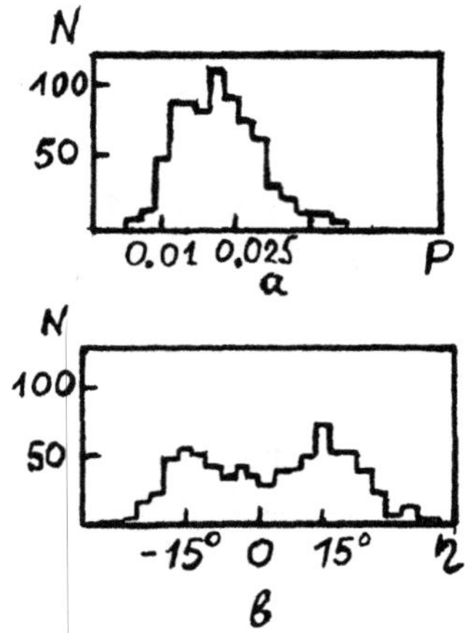

Figure 69. Histograms showing the distribution of the degree of linear polarization P of D_3 helium line (a), and of the deviation η of the position angle of the oscillating electric vector, with respect to the tangential direction in a quiet prominence (b) according to /94/

the angles ξ and ψ (Fig. 68) were determined by comparing the observed values of the degree of polarization P, the polarization direction (specified by the angle η ; Fig. 68) and the spectral line intensity I_0 with the results of a quantum-mechanical calculation for the Hanle effect for the line of interest in a prominence. This calculation was based on a model of free atoms excited by resonance radiation from the photosphere in the absence of a local magnetic field. The probability of collisional excitation and depolarization of the line of interest was assumed to be negligible, since collisional processes are important only deeper in the solar atmosphere. The polarization tensor of the exciting light field was calculated for each point in the prominence, and then the polarization characteristics of the helium D_3 line light scattered in the observation direction were calculated numerically, giving the emission intensity I_0, the degree of polarization P, and the rotation angle η of the predominant polarization direction, as a function of the altitude of the observation point, the density of helium atoms, the electron density and temperature, and the magnetic field vector $B(\xi, \psi)$. These calculations yielded polarization diagrams: families of curves corresponding to the dependence of P/P_{max} and η on the local magnetic field $B(\xi, \psi)$. The calculations were carried out by V. Bommier for fields both weaker and stronger than 15 Gauss /96/. In stronger magnetic fields, the polarization diagrams became multivalued as a result of crossings and anticrossings of

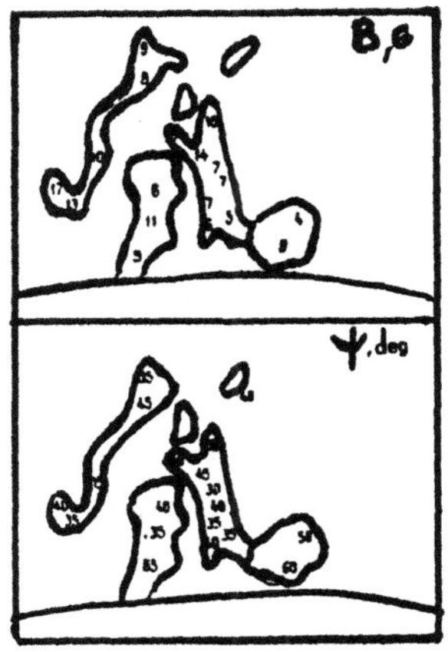

Figure 70. Map of the magnetic field of a quiet prominence obtained in /97/ through an analysis of the Hanle effect

the Zeeman sublevels. This study yielded maps of the magnetic field in quiet prominences according to polarimetric measurements of the helium D_3 line. Fig. 70 shows an illustrative magnetic-field map of a prominence observed on 19 December 1974.

The measurements of an arbitrarily oriented magnetic field vector can be made more reliable by simultaneously observing the polarization characteristics of the radiation in two helium spectral lines, at 5876 Å and 10 830 Å. In those studies it was assumed that the high optical transparency of the medium would make multiple scattering of photons unlikely. On the other hand, an increase in the density of the medium generally leads to absorption of the radiation, and even a few photon reabsorption events can cause a significant additional depolarization of the radiation. This question was taken up theoretically by House and Cohen /98/, who applied the Monte-Carlo method to a simplified model of a quiet prominence.

7.1.3 CORONAL MAGNETIC FIELD

In the solar corona one observes forbidden lines corresponding to magnetic-dipole transitions in multiply charged iron and calcium ions. A detailed analysis of the conditions for the excitation of the Fe XIV 5303 Å and Fe XIII 10747 Å lines showed that the anisotropy of the excitation of these lines and thus the nature of the polarization of the scattered light are determined to a large extent by photoexcitation by photospheric light. Charvin

/99/ carried out detailed calculations of the polarization of the radiation in the green Fe XIV 5303 Å line for several models of the corona and for several simple magnetic-field configurations; radiative and collisional excitation and the local magnetic field were taken into account. Expressions were derived for an upper limit of the degree of polarization as a function of the distance from the edge of the disk, and it was shown that, under favorable conditions, measurements of the polarization could furnish in the first place information on the direction of the projection of the local magnetic field onto the plane perpendicular to the observation line, and secondly the electron density in the plasma. These measurements become more reliable when simultaneous use is made of several spectral lines. The possibilities of magnetometry of the solar corona on the basis of polarimetric measurements were examined in a number of studies of the polarization of the radiation from an ensemble of particles in the case of anisotropic excitation by resonance radiation from the photosphere. Within a corona model of nonuniform temperature and density, with an arbitrary spatial distribution of the magnetic field, the degree of linear polarization and the rotation of the polarization plane were calculated with allowance for integration along the observation line. The results are applicable to both electric and magnetic-dipole transitions.

Hyder's analysis /100/ of the polarization studies of solar corona lines was based on Hyder and other observers' measurements. The measurements were made for several spectral lines of multiply charged ions, Fe X to Fe XV and Ca XV. Assuming an optical mechanism for the alignment, the degree of polarization of the radiation from an ensemble of particles was calculated as a function of the distance to the solar surface. The resulting theoretical values were compared with measurements for lines with known g-factors and for transitions with upper levels of known lifetimes. As a result, it was found that the range of magnetic fields in the solar corona is $5 \cdot 10^{-8} \leq B \leq 5 \cdot 10^{-9}$ Gauss. Detailed studies of the green Fe XIV 5303 Å and the red Fe XV 6374 Å lines of the solar corona /101/ have shown that the degree of polarization of the green line varies over the range 2-2.5% for $1.0 \leq R/R_{\odot} \leq 1.5$, while that for the red line is at the level of $2 \pm 3.5\%$, in agreement with the calculations of /99, 100/. The polarization of the Fe XV and Fe XIII coronal lines was studied theoretically in a number of papers, where the role played by depolarizing collisions of protons was taken into account along with alignment in the photoexcitation of ions. The polarization characteristics of the green line of Fe XIV were analyzed in /102/ under the condition $\omega\tau^{(2)} \gg 1$, i.e., in the absence of a Zeeman coherence in the system. In this case, the ensemble of particles is characterized exclusively by the population and the longitudinal alignment, while the radiation from the ensemble is characterized by the intensity and degree of linear polarization, which depend on the orientation of the magnetic field, the temperature, the electron and ion densities, the degree of ionization, and other parameters. The theoretical investigation of the polarization of the ultraviolet Si IV 1394 Å and Ca III 977 Å lines in various parts of the chromosphere and the corona, in a model of coherent optical excitation in the absence of a magnetic field, was carried out /103, 104/. Then the effect of a weak magnetic field, and of collisions in an optically dense medium, on the polarization of these emission lines was studied /105/.

7.2 Impact polarization in solar flares

A rich body of literature exists on the explosive manifestation of energy release that takes place during solar flares /106, 107, 108, 109/, since solar flares are among the most interesting phenomena in astrophysics both because of the physics involved and because of their impact on the earth. They are also an example of quantitative application of spectropolarimetry for sensing the energy transport from the corona, where the main energy release takes place, to the upper chromospheric layers. The anisotropies associated with the various possible mechanisms of energy transport lead to linear polarization of line radiation, that is used as a diagnostic tool to determine the characteristics of the primary energy release process.

7.2.1 SOLAR FLARES

Solar flares are localized, short-duration (1 minute to hours) increases of the solar radiation field, ranging from γ-rays to kilometric-radio-wavelengths, associated with particle acceleration and mass motions. Solar flares are dominantly observed in regions of strong and complex magnetic fields. Even if not completely observationally proved, there is wide agreement that the energy released in solar flares is provided by the dissipation of magnetic field energy. Chromospheric solar flares have been observed for decades as strong brightenings of the H_α chromospheric line, whereas the presence of accelerated particles in solar flares was discovered with the development of radioastronomy and of X-ray astronomy. The acceleration region is usually assumed to be localized in the corona, above the chromosphere, in regions where complex magnetic fields are present. Local magnetic energy release is not entirely excluded. However, chromospheric flares are usually considered to be secondary phenomena (Fig. 71). Magnetic energy is assumed to be dissipated mainly in the corona, producing enhanced heating and particle acceleration. Then the energy released can be transferred from the corona to the chromosphere.

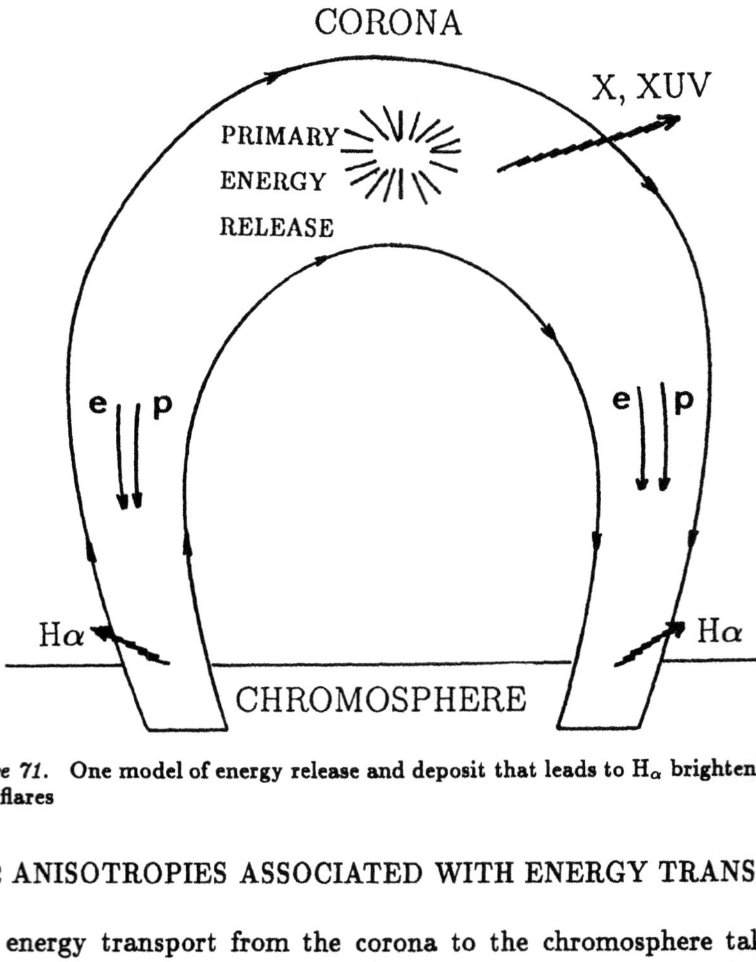

Figure 71. **One model of energy release and deposit that leads to H_α brightening during solar flares**

7.2.2 ANISOTROPIES ASSOCIATED WITH ENERGY TRANSPORT

The energy transport from the corona to the chromosphere takes place via heat conduction, particle bombardment or X-ray irradiation. All of these mechanisms of energy transport involve electrons or protons with anisotropic velocity-distribution-functions:

- the velocity-distribution function of electron or proton beams is peaked in the magnetic field direction;

- the velocity-distribution function of electrons carrying heat flux has a tail in the heat flux direction;

- photoelectrons produced by irradiation of the solar chromosphere by an XUV coronal radiation source are predominantly moving in the solar horizontal plane.

Measurements of the linear polarization of a chromospheric line, strongly enhanced during solar flares, impart information on the relative importance of the various modes of energy transport. The association between conduction, particle bombardment, X-ray irradiation and the direction of polarization of collisionally excited H_α radiation is illustrated in Fig. 72 and discussed below.

- Assuming that the average magnetic field that guides the beam of highly energetic electrons (of energy higher than 20 keV) is vertical, it can easily be shown that, observed on the solar disk, the main direction of the polarized electric vector of the collisionally excited H_α line is perpendicular to the flare-to-disk-center direction.

- Photoelectrons have an energy close to the H_α excitation threshold and produce a polarization fraction opposite to the one caused by a beam of energetic keV electrons having the same direction of motion. Since photoelectrons are ejected perpendicular to the direction of motion of the incoming photon, a compact XUV radiation source above the chromosphere will generate photoelectrons whose velocity-distribution function will be maximum in the horizontal plane. Therefore, they could lead to the same direction of polarization of H_α line radiation as a vertical beam of keV electrons.

- Heat flux is carried by the electrons with a few times the thermal velocity. At the bottom of the corona-chromosphere transition zone, they have an energy close to the H_α excitation threshold and a vertical direction of propogation. As a result they would lead, by collisional excitation, to the emission of polarized H_α radiation with an electric vector in the flare-to-disk-center direction.

- High energy protons above 10 MeV have been observed in solar flares. Nevertheless, from observations, nothing is known about energetic protons below 1 MeV. Beams of deca-keV protons falling along the vertical magnetic field and producing hot plasma have been recently suggested for explaining X-ray emission. Such beams of protons would also produce H_α line polarization with a polarization vector in the flare-to-disk-center direction.

7.2.3 H_α POLARIZATION OBSERVATIONS

In 1981, during a solar flare, linear polarization was observed from space using the Ultraviolet Spectro-Polarimeter on the Solar Maximum Mission Satellite (SMM) in the integrated profile of the 1437 Å line emitted in the solar chromosphere /110/. At that time ground-based spectropolarimetric solar flares observations were begun at Meudon Observatory in the hydrogen Balmer H_α line. These observations were made using an H_α flare patrol heliograph /111/. This instrument provided monochromatic images of solar active regions in the $H\alpha$ line center with a wavelength band-pass of 0.75 Å. The polarimetric analysis was made by rotating a half-wave plate inserted in front of the linear polarizer at the entrance of the Lyot filter, that isolates the center of the $H\alpha$ line. By rotating the half-wave plate through

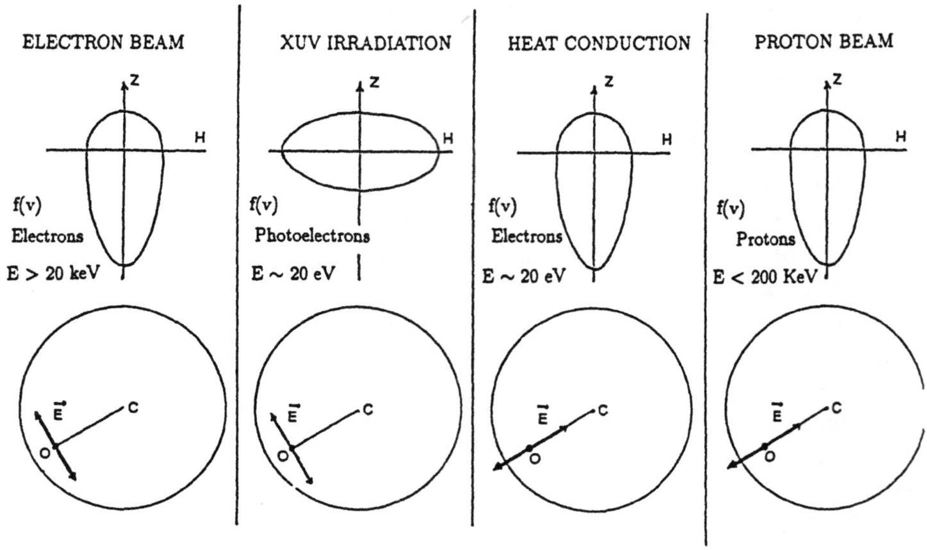

Figure 72. **Direction of vibration of the polarized electric vector, projected on the solar disc for four different energy transport mechanisms**

22.5 degrees per step, a polarization analysis can be carried out, leading to the polarization degree and orientation of the linear polarization. On the photographic first version of the instrument, sixteen H_α filtergrams were obtained on photographic film every minute.

The intensity in each point of the filtergrams is a function of the H_α linear polarization and of the position of the half-wave plate. After digitizing the filtergrams, the relative positioning of the successive images can be made by a least-square fit computer code. The Stokes parameters P and Q are then derived by adding and substracting the appropriate filtergrams. Four independent determinations of the Stokes parameters per minute can be obtained.

A few chromospheric flares were observed in July 1982. These flares took place in a complex active region located at 16 degrees N, 50 degrees E. By integrating over 10 minutes of observing time, a linear polarization fraction of a few percent was observed at some locations in the bright H_α patches.

The histograms of Fig. 73 show the distribution of the orientations of the polarized electric vector obtained in two flaring portions of an active region observed during a flare on 17 July 1982 /111/. The mean polarization was obtained by integrating in time over about ten minutes and in space over an area of about 2000 $\times$ 2000 km^2. A peak in the histograms is present

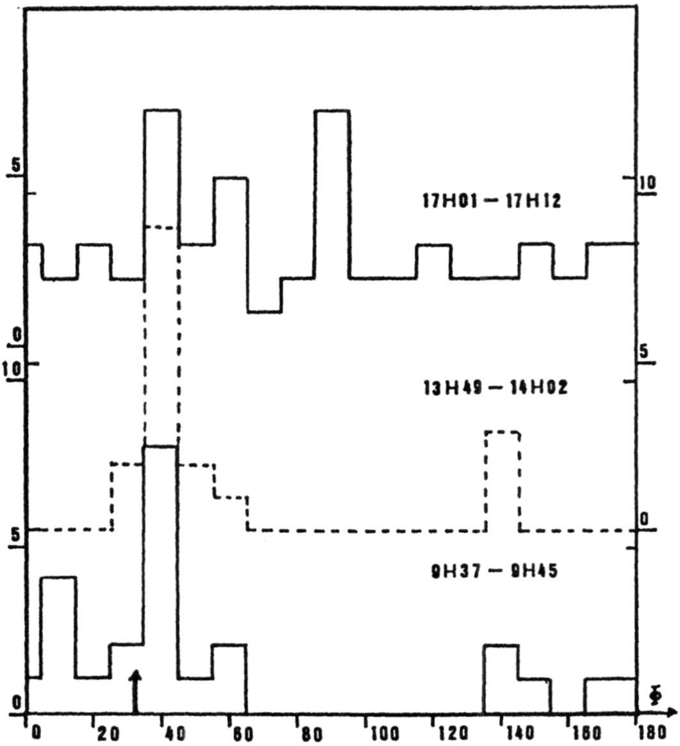

Figure 73. Number of pixels, where the observed H_α polarization fraction, integrated over about 10 mn, is greater than 2.5 %, in three flares as a function of the orientation angle between the polarized electric field vector and an axis of reference. An arrow indicates the mean value of the orientation angle of the flare-to-disk-center direction for these flares

for a direction of the polarized electric vector that differs from the flare to disk center direction by no more than 10 degrees. Then Figs. 74, 75, 76 show the location on the filtergrams of the pixels where this polarization was present during flares observed on 11 July 1982 /112/. In 1989, the variation of the polarization degree over the H_α line profile during solar flares was observed using the large telescope of Baikal Astrophysical Observatory of the Russian Institut for Solar and Terrestrial Physics (RISTP) /39,113/. Spectropolarimetric measurements were made using two crossed Wollaston prisms mounted behind the entrance slit of the spectrograph. The relative intensities of the two orthogonally polarized ordinary and extraordinary beams respectively, parallel and perpendicular to the grating ruling, were equalized by rotating an achromatic $\lambda/2$ plate behind the Wollaston prisms. Another plate was mounted before the slit in order to determine the azimuth of the polarization plane.

Figure 74. Digitized H_α filtergrams showing H_α chromospheric brightening during three solar flares. The $H\alpha$ brightness is represented by isocontours - respectively in full and dotted lines - for intensities above or below the mean intensity in the total field of view. The direction of the polarized electric vector in the flaring regions, together with the flare-to-disc-center direction, is shown by arrows. The integration time in the first flare is 8 mn

Figure 77 shows a drawing of the image of an observed active region in the plane of the entrance slit averaged over the observation time. Positions of the entrance slit at different moments are noted by numbers 1-4. The observations show that the central part of H_α emission line is linearly polarized. As an example of the observational results, the spectral profiles of two bright kernels of a relatively strong flare are given in the Fig. 78a,b,c,d. For the first node in the center of $H\alpha$ line the degree of the linear polarization was about 4% (no polarization was registered in the second node).

7.2.4 ORIGIN OF THE OBSERVED POLARIZATION

At first sight the observed polarization could be due either to Zeeman effect, Stark effect, or to impact by particle beams. Zeeman polarization would be located above sunspots and the direction of the polarized electric vector

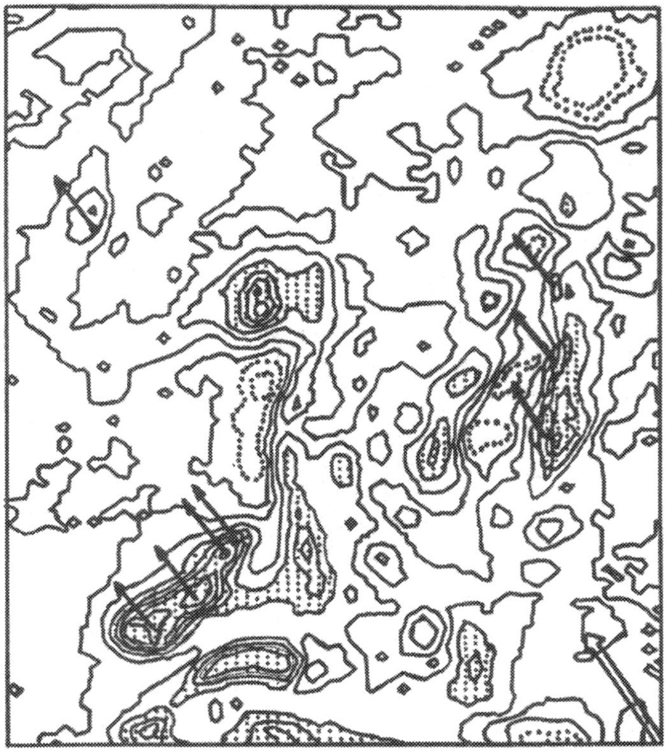

Figure 75. Second flare in the same active region as in Fig.72 with an integration time of 13 mn

would reproduce the fan-shape of the magnetic field lines. That is not observed. The Doppler broadening of the H_α line leads to a negligible Zeeman polarization at line center (less than 0.1 %). Turbulent electric fields originating from plasma turbulence may produce linear polarization by the Stark effect. In the case of strong turbulence, the turbulent field reaches 170 V/cm for the chromospheric flare conditions. However, the corresponding Stark spacing produced by the dynamical electric field is only $5\,10^{-4}$ nm, much smaller than the 0.07 nm Doppler width. Then the polarization fraction of the Hα radiation integrated over the Lyot filter band-pass, that could be produced by the Stark effect, is negligible. Consequently the observed polarization can be interpreted as impact polarization.

The vibration direction of the polarized electric vector is close to the flare-to-disk-center direction. This gives information on the dominant direction of propagation of the impacting particles. Fig. 79 shows the observed and computed energy dependence of the polarization fraction of the H_α line at 90^0 of an electron beam. The polarization degree at 90^0 of the beam reverses at an energy close to 50 eV. Less information is available for pro-

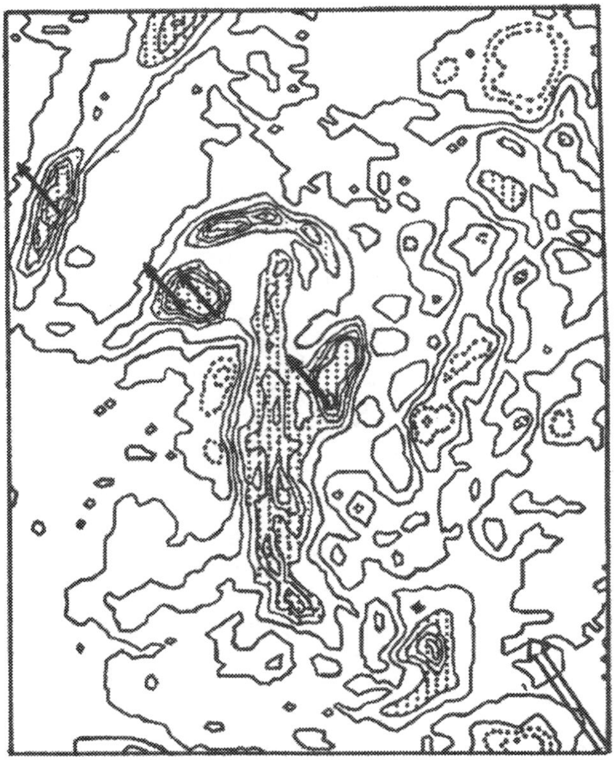

Figure 76. Third flare in the same active region as in Fig.72 with an integration time of 11 mn

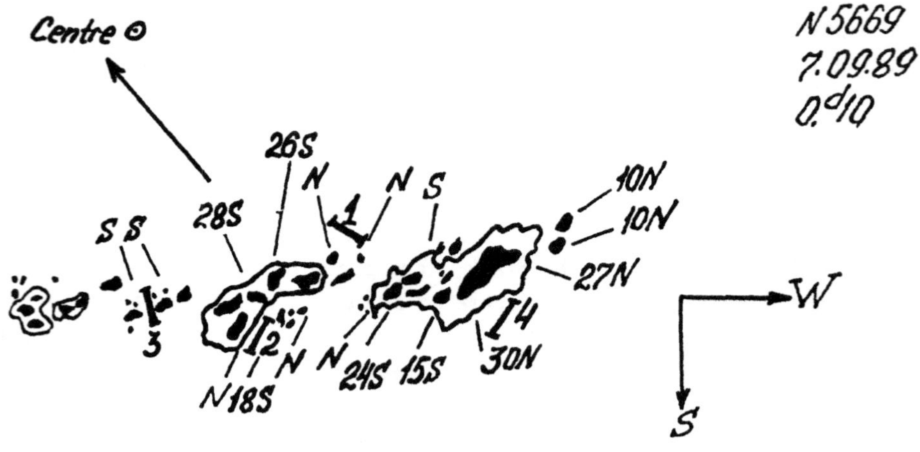

Figure 77. Successive spectrograph slit positions (1 to 4) with respect to a group of spots on 07.09.1989 at different moments

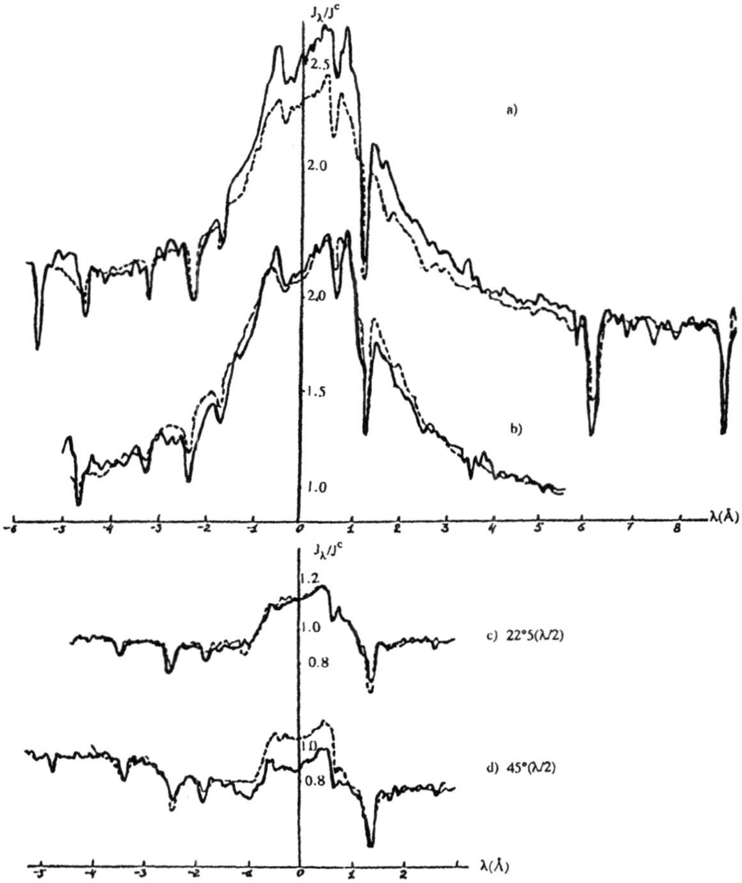

Figure 78. **a and b : H_α profiles of a bright H_α flare kernel for two positions of the entrance half-wave plate; c and d : profiles obtained 1 minute later for the same conditions of polarization analysis.**

tons. However, a similar behaviour is observed for beams of protons with a reversal at higher energy (about 100 keV). The polarization degree below this reversal energy is positive, which means that at 90^0 of the beam the polarization vector is parallel to the beam direction. This implies that the polarization is produced either by low energy particles (energy lower than the reversal energy) moving predominently in a direction parallel to the solar vertical direction, or to high energy particles circling in the horizontal plane. Since the Hα line excitation cross-section decreases with energy, the second hypothesis would require both a higher number of particles and a higher energy flux to explain the observed polarization. Therefore, impact with neutral hydrogen of low energy particles - deca-eV electrons or deca-

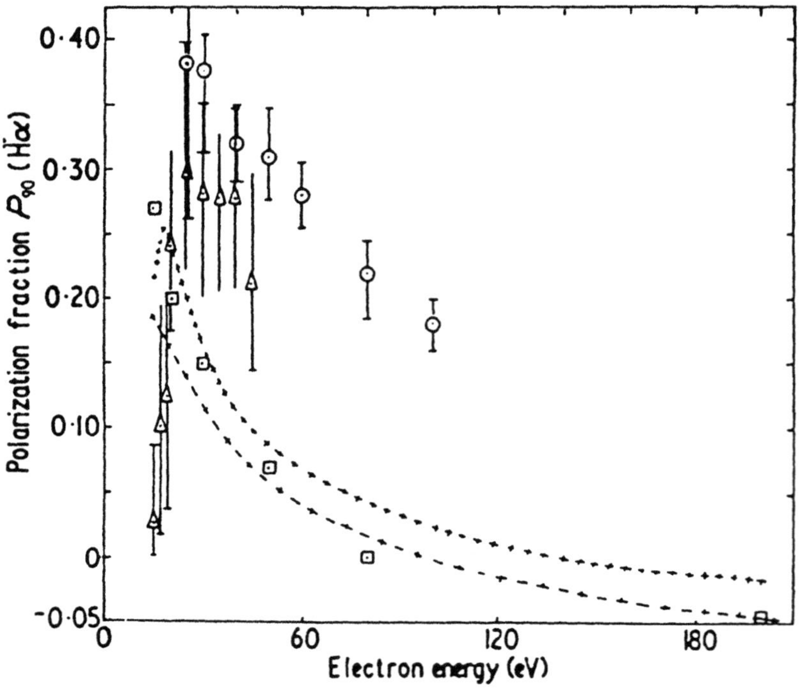

Figure 79. Observed and computed values of the H_α line polarization fraction at 90° to an electron beam as a function of impact energy /114/. Pluses and alternate plus and minus signs represent respectively DWPO II and Born theoretical values. Circles and triangles are experimental values /115, 116/. Squares are theoretical values calibrated with measured integrated cross sections /117/.

keV protons - moving vertically is the most likely origin of the observed polarization.

Knowing that deca-eV electrons and deca-keV protons presumably come from the corona and bombard the chromosphere as suggested in Fig. 71, the choice between these particles requires one to examine the few possibilities of accelerating them in the solar vertical direction, without excluding the possibility of local acceleration /38/.

1 - Polarization can be generated in double layers. A double layer is a localised voltage drop in an electric current path that could occur when the local density of electric current carriers is too low for carrying an increasing electric current density, and consequently charge of neutrality cannot be maintained locally. The formation of strong double layers with a high voltage drop requires a current density presumably too high to be present on the sun. However, weak double layers, where the energy of the poten-

tial drop is of the order of the particle thermal energy, are more easily generated. These double layers would locally accelerate particles, producing electrons of a few tens of eV with an anisotropic velocity distribution function. These electrons would excite the upper level of Hα transition and produce the polarized line emission. Indeed, this hypothesis raises the question of the possibilty of existence of the very high current density that is required for the formation of a double layer. This last condition implies that the electric currents in which the flare magnetic energy is stored are flowing in structures less than 10 km wide. Since these structures cannot be resolved by existing solar telescopes, the question remains open.

2 - Strong temperature gradients can produce a departure from an isotropic Maxwellian distribution. Attempts have been made to relate the observed linear polarization to the heat flux in the high solar chromosphere. The polarisation fraction is related to the even coefficients of the Legendre polynomial expansion of the velocity-distribution-function, when the conductive flux is related to the odd coefficients. Therefore, the relationship between these coefficients must be found before deducing the heat flux from polarization measurements. The derivation of such an exact relationship requires one to solve a Fokker-Planck equation in the solar transition zone between the corona and the chromosphere for electrons of energy close to the thermal energy of the transition region. This is a difficult task and has not yet been achieved. Moreover, existing computations suggest that since the electron number density in the chromosphere is high (10^{11} -10^{13} cm^{-3}), low energy electrons of a few tens of eV are isotropised within very short distances.

3 - Proton beams could also lead to linearly polarized line emission. Contrary to electrons, heavy protons keep to a more or less straight line path while losing energy. The most efficient protons would have an energy between 20 and 100 keV. Deca-keV protons accelerated in the corona could go through a significant mass of plasma at high coronal temperature (10^7 K) without momentum exchange. As long as this coronal mass is not too high, they will reach the high chromosphere with their initial directivity. Contrary to deca-eV electrons, their local production in the high chromosphere by a potential drop is unlikely, requiring the improbable formation of strong double-layers.

On the one hand, the particular location of the high chromospheric layers emitting Hα core, close to the region of the high temperature gradient, is in favor of deca-eV electrons coming from the transition zone. But exact computation of the characteristics of the electron velocity distribution function at the bottom of the transition zone is required to check the validity of this interpretation. Most existing computations are valid for the transition region and for rather energetic electrons, and the chromosphere where we need to estimate the remaining anisotropy is most often considered for simplicity as a boundary layer with an isotropic velocity distribution function of thermal electrons. Acceleration in weak double layers is also a possible interpretation. On the other hand, deca-keV proton beams accelerated in the corona and reaching the high chromosphere with a still significant directivity are good candidates to offer an explanation for the observed polarization.

Another source of energetic electrons with an anisotropic velocity-distribution-function in the chromosphere is photo-ionizing XUV radiation emitted by discrete sources in the transition region and corona. However, it would lead to a polarization perpendicular to the radial direction.

7.3 Energy transport by protons in solar flares

Discussing the possibility for proton beams accelerated higher in the atmosphere to explain the observed polarization /38, 112/, we shall successively examine the two following conditions that have to be satisfied for these protons:
1. Their velocity distribution function must have a significant anisotropy at chromospheric levels.
2. Their flux has to be high enough for the collisional excitation rate of hydrogen by these protons to exceed the depolarizing collision rate by the background electrons.

7.3.1 EXPECTED PROTON VELOCITY ANISOTROPY

In the absence of scattering, the equation of conservation of the proton number flux for a beam propagating in a magnetic loop of section s, with a pitch angle (angle between the proton beam velocity and the magnetic field vectors) α is:

$$s_0 f_0(p_0, \alpha_0) p_0^3 dp_0 \sin\alpha_0 \cos\alpha_0 d\alpha_0 = s f(p, \alpha) p^3 dp \sin\alpha \cos\alpha d\alpha \qquad (7.1)$$

where $p = m_p v$ and the subscript 0 is used to represent the conditions at the top of the transition region. The conservation of the adiabatic invariant and of the magnetic flux, i.e.

$$Bs = B_0 s_0 \quad \text{and} \quad \sin^2\alpha/B = \sin^2\alpha_0/B_0,$$

together with equation (7.1), lead to

$$f_0(p_0, \alpha_0) p_0^3 dp_0 = f(p, \alpha) p^3 dp \qquad (7.2)$$

The loss of angular momentum of a beam is given by /118/ :

$$p_0^4 - p^4 = 4\,m_p^2 \frac{K N_b}{\mu} \tag{7.3}$$

where $K = 4\pi e^4 \Lambda\, m_p/m_e$, $\mu = \cos\alpha$ and N_b is the column density of the region above the Hα forming layers where the magnetic field is strong enough to rule out scattering by magnetohydrodynamic waves. It is reasonable to take for its value the column density of the chromosphere above these layers. Using relation (7.3), equation (7.2) becomes:

$$f(p,\alpha) = f_0(p_0, \alpha_0) \tag{7.4}$$

Equation (7.4) shows that for an initial isotropic velocity distribution the magnetic field convergence does not affect the anisotropy created by atmospheric stratification. As it is usually assumed, at the top of the transition region the initial velocity distribution is isotropic and the energy dependence of the particle flux is represented by a power law $E^{-\delta}$ i.e. $f_0(p_0, \alpha_0) \propto p_0^{-2\gamma}$, where $\gamma = \delta + 1$. Thus equation (7.4) gives

$$f(p,\alpha) \propto \left(p^4 + 4\frac{K\,N_b\,m_p^2}{\mu}\right)^{-\gamma/2}.$$

At energies close to the excitation threshold $p_{th}^4 \ll 4KN_b m_p^2/\mu$ and

$$f(p_{th},\alpha) \propto (4KN_b m_p^2)^{-\gamma/2}\,\mu^{\gamma/2}. \tag{7.5}$$

This velocity distribution is anisotropic and could therefore lead to the local emission of the linearly polarized H-alpha photons.

For a line of sight making an angle θ with the local solar vertical, the polarization $\mathcal{P}(\theta)$ is defined:

$$\mathcal{P}(\theta) = \mathcal{P}(90°)\sin^2\theta/(1 - \mathcal{P}(90°)\cos^2\theta)$$

where $\mathcal{P}(90°)$ is given by

$$\mathcal{P}(90°) = \frac{\int_{v_0}^{\infty} P(90°,v)[(3J_2 - J_0)/(3 - P(90°,v))]v\sigma(v)\,dv}{\int_{v_0}^{\infty}\{[2J_0 + P(90°,v)(J_2 - J_0)]/(3 - P(90°,v))\}v\sigma(v)\,dv}.$$

$\sigma(v)$ is the total collisional excitation cross section, v_0 is the incident particle velocity at threshold.

$$J_n = \int_0^{\pi} f(v,\alpha)\cos^n\alpha\sin\alpha d\alpha$$

is the nth moment of the particle velocity-distribution-function $f(v,\alpha)$, where α is the angle between the particle velocity vector and the solar vertical. By introducing an effective velocity $\overline{v}$, the polarization can be expressed as

$$\mathcal{P} = P(90^\circ)\sin^2\theta\, b(\overline{v}),$$

where $b(\overline{v})$ is defined as

$$b(\overline{v}) = [3J_2(\overline{v}) - J_0(\overline{v})]/2J_0(\overline{v}).$$

By expressing the velocity distribution function as a sum of Legendre polynomials:

$$f(v,\alpha) = \sum_{\kappa} f^{\kappa}(v)P^{\kappa}(\cos\alpha),$$

the coefficient $b(\overline{v})$ is found to be equal to $f^2(\overline{v})/5f^0(\overline{v})$. The maximun polarization that can be observed at 90 degrees to the solar vertical is given by:

$$\mathcal{P}(\theta) = P(90^\circ,\overline{v})\, b(\overline{v}),$$

showing that the polarization fraction depends on the energy transport mechanism via the ratio of the coefficients and $f^2(\overline{v})$ and $f^0(\overline{v})$ in the Legendre polynomial expansion of the associated particle velocity distribution function at some effective velocity $\overline{v}$.

The anisotropy factor of the velocity-distribution function given by eq. (7.5) is found to be $b(v_{th}) = \gamma/(\gamma+6)$. For $\gamma = 5$ $(\delta = 4)$, $b(v_{th}) = 0.45$ and the difference in energy loss for protons travelling with different inclinations on the solar vertical leads to a degree of linear polarization at the limb of the sun as high as 13%. A near isotropy of the proton velocity distribution at the top of the transition region - due either to the injection process or to scattering by waves in the corona - does not preclude an anisotropic proton velocity distribution at the H_α forming layer.

The proton number flux in the chromosphere has to be sufficiently high in order to produce linearly polarized H_α radiation. The H_α line center forming layer is at a temperature $T \geq 9 \times 10^3$ K. Above this layer almost all proton energy loss occurs in collisions with free electrons of the highly ionized plasma. In Coulomb collisions with the background electrons, only electrons of velocity smaller than the velocity of the 100 keV protons, i.e. electrons of energy below a few 100 eV, contribute to the energy losses. In the corona their number density is a small fraction of the total electron number density, and protons can go through a significant mass of coronal plasma (hot target) without momentum exchange. This is no longer true in the cooler atmosphere below (cold target).

7.3.2 DELIVERED ENERGY FLUX

For a proton moving with constant pitch angle α in a homogeneous vertical magnetic field, the energy loss per unit electron column density under collisions with free electrons is:

$$\frac{dE}{dN} = -\frac{2\pi e^4 \Lambda}{\mu E}\left[\frac{m_p}{m_e}\varphi(x) - \varphi'(x)\right] \tag{7.6}$$

where $\mu = \cos\alpha$, Λ is the Coulomb logarithm, $x = (E/kT)(m_e/m_p)$, and $\varphi(x) = \Phi(\sqrt{x}) - \sqrt{x}\Phi'(\sqrt{x})$. $\Phi(x)$ is the error function.

In the chromosphere, one can always use the cold target approximation, $x \gg 1, \varphi(x) \approx 1$. In the corona, on the other hand, the hot target approximation, $x \ll 1, \varphi(x) \approx 4/3x^{3/2}\pi^{-1/2}$, may be applied. However, the following expression is a better approximation of $\varphi(x)$ valid for all x :

$$\varphi(x) = \left(1 + \frac{3\sqrt{\pi}}{4x^{3/2}}\right)^{-1} \tag{7.7}$$

Let N_a and N_b be the total column densities of the corona and chromosphere respectively, T_a the mean temperature of the corona and E_0, E_1 and E, the successive energies of a proton at injection, at the entrance of the chromosphere and in the H_α line center forming region. Neglecting the $\varphi'(x)$ term, integration of equation (7.6) with the help of equation (7.7) for the corona and of the cold target approximation for the chromosphere gives the following relations:

$$E_0^2 - E_1^2 + 3\sqrt{\pi}U_a^{3/2}(E_0^{1/2} - E_1^{1/2}) = KN_a/\mu \tag{7.8}$$

$$E_1^2 - E^2 = KN_b/\mu \tag{7.9}$$

where we have set $K = 4\pi e^4 \Lambda m_p/m_e$ and $U_a = KT_a m_p/m_e$. Thus the minimum energies E_b and E_{ab} required, respectively, to cross the chromosphere and the whole atmosphere above the Hα forming layer are given by the equations

$$E_b^2 = KN_b/\mu \tag{7.10}$$

$$E_{ab}^2 - E_b^2 + 3\sqrt{\pi}U_a^{3/2}(E_{ab}^{1/2} - E_b^{1/2}) = KN_a/\mu. \tag{7.11}$$

With E_b known from the equation (7.10), the equation (7.11) gives E_{ab} versus T_a.The constraint on the initial flux is found by equating the collisional excitation rate of hydrogen level 3 by the beam C_{13}^B to the thermal excitation rate by the background electrons C_{13}^T . The beam collisional excitation rate is

$$C_{13}^B = \int_0^\infty n^B(E)\nu(E)\sigma_p(E)dE,$$

where $n^B(E)$ is the local beam particle density at energy E and $\sigma_p(E)$ is the cross-section for excitation by proton impact. Strictly speaking, the lower limit of the integral should be the excitation energy threshold, but in practice it makes little difference to take it as zero. One can simply estimate $\sigma_p(E)$ by noting that about one-tenth of the total energy lost by a proton in colliding with neutral hydrogen is spent to excite level 3. The energy loss rate /112/ is

$$(\frac{dE}{dt})_H = -\frac{2\pi e^4 \Lambda'}{E}\frac{m_p}{m_e} n_H\, v,$$

where Λ' is an "effective Coulomb logarithm". It follows that

$$\sigma_p(E) = \frac{S}{E}$$

where $S = 0.2\pi e^4 \Lambda' \chi_{13}^{-1} m_p/m_e = 2.8 \times 10^{-15}$ keV cm^2, $\Lambda' = 1.4$ and the excitation potential of the hydrogen level 3, $\chi_{13} = 12.1$ eV. Thus

$$C_{13}^B = S \int_0^\infty \frac{n^B(E)v(E)dE}{E},$$

which can be transformed into an integral over E_0 by writing the flux conservation condition

$$n^B(E)v(E)dE = n_0^B(E_0)v(E_0)dE_0$$

and solving equations (7.8) and (7.9) in order to express E as a function of E_0. This is difficult, but a simple estimation may be obtained as follows. For a power-law spectrum at injection,

$$n_0^B(E_0)v(E_0) = \eta E_0^{-\delta},$$

the main contribution will be from protons of energy not much larger than E_{ab}. Thus, using the expansions

$$E_0^2 = E_{ab}^2(1 + \varepsilon_0)$$

and

$$E_1^2 = E_b^2(1 + \varepsilon_1)$$

one obtains $\epsilon_1 = \lambda\epsilon_0$ and $\epsilon_1 = E^2/E_b^2$, where

$$\lambda = \frac{4E_{ab}^2 + 3\sqrt{\pi} U_a^{3/2} E_{ab}^{1/2}}{4E_b^2 + 3\sqrt{\pi} U_a^{3/2} E_b^{1/2}},$$

and equation (7.12) becomes

$$C_{13}^B = \eta S \int_{E_{ab}}^{\infty} \frac{E_0^{-\delta} dE_0}{\lambda^{1/2} E_b (E_0^2/E_{ab}^2 - 1)^{1/2}},$$

or

$$C_{13}^B = \frac{\eta S}{2} \frac{E_{ab}^{1-\delta}}{\lambda^{1/2} E_b} \int_0^1 \frac{u^{\delta/2-1}}{(1-u)^{1/2}} du.$$

The remaining integral is a Beta function, so one finally obtains, in terms of

$$F_{ab} = \eta \mu E_{ab}^{2-\delta}/(\delta - 2),$$

the total flux of energy carried by protons injected at $E_0 > E_{ab}$:

$$C_{13}^B = S \frac{\delta - 2}{2\mu} B(1/2, \delta/2) \frac{F_{ab}}{\lambda^{1/2} E_{ab} E_b}.$$

The thermal excitation rate is

$$C_{13}^T = \int_{v_0}^{\infty} v f(\mathbf{v}) \sigma(v) d^3 v,$$

where

$$v_0 = (2\chi_{13}/m_e)^{1/2}.$$

The velocity vector is $\mathbf{v}$ and $f(\mathbf{v})$ is the local Maxwellian velocity distribution function for an electron number density n_e and a temperature T. Since in the Hα forming region $\chi_{13} \gg kT, f(v)$ decreases quickly with velocity and the excitation cross-section, $\sigma(v)$, can be taken equal to its value $\overline{\sigma}$ near threshold. Thus with $T = 9000$ K and $\overline{\sigma} = 5 \times 10^{-17}$ cm^2 one gets

$$C_{13}^T \approx 7.7 \times 10^{-15} n_e.$$

For a flux of protons $\propto E^{-\delta} dE$ with $\delta = 4$, the minimum energy flux F_{ab} above E_{ab} that is required in order to satisfy the condition $C_{13}^B = C_{13}^T$ is given by

$$F_{ab} = 3.33 \mu 10^{-9} n_e \lambda^{1/2} E_{ab} E_b$$

where F_{ab} is in ergs cm^2 s^{-1} and E_{ab} and E_b are in keV.

The energy requirements depend strongly on the coronal total column density and on the coronal temperature. In the solar flare model F_1 /119/ for the temperatures characteristics of the flaring coronal plasma ($T_a \geq 10^7$ K), F_{ab} does not exceed 3x10^8 ergs cm^{-2} s^{-1}. For $T_a = 10^7$ K, $E_{ab} =$ 200 keV and the associated total photon number flux is equal to 5x10^{14} particles cm^{-2} s^{-1}. Again the overestimated coronal total column density of the model F_2 /119/ leads to somewhat unrealistic energy requirements.

Extrapolation of the spectrum to a standard energy cut-off E_e would lead to an extrapolated energy flux

$$F_e = F_{ab}\left(\frac{E_{ab}}{E_e}\right)^2$$

The results presented above were obtained for an unidirectional beam of zero pitch angle. For another angular distribution of velocities it would be necessary to integrate over the pitch angles. However for a nearly isotropic distribution it is sufficient to use the expressions above with an average pitch angle of 60 degrees. In this case the energy flux has to be multiplied by a factor of 2-3.

7.3.3 APPLICATION OF THE POLARIZATION MOMENT FORMALISM

Assuming that the polarization of H_α hydrogen line (n=3) is induced by collisions with precipitating particles having a preferential direction in the velocity space, we shall describe the ordering of the moments of the excited hydrogen atoms with the help of the polarization moments formalism $\rho_q^{(x)}(ll')$ (Chapter 1) independently of the nature of the projectiles /120, 121/:

$$\rho_q^{(\kappa)}(ll') = \sum_{m,m'} (-1)^{l'-m'} \left[\begin{matrix} l & l' & x \\ m & -m' & q \end{matrix} \right] \rho_{mm'}^{ll'}, \tag{7.12}$$

where $\rho_{mm'}^{ll'}$ are the density matrix elements in the representation of the magnetic quantum numbers. $\left[\begin{matrix} l & l' & x \\ m & -m' & q \end{matrix} \right]$ is the Clebsh-Gordan symbol.

Let us introduce two frames of reference (Fig. 80) $\tilde{x}, \tilde{y}, \tilde{z}$ with $O\tilde{Z}$ axis directed along the symmetry axis of the collisional process and the system $\approx{x}, \approx{y}, \approx{z}$ with $O\approx{Z}$ axis turned at β angle with respect to $O\tilde{Z}$ and oriented along the relative velocity vector between the two colliding particles.

As the collision process in the frame of reference $\approx{x}, \approx{y}, \approx{z}$ is axially symmetric, the non-zero polarization moments in this system will have only zero q-numbers $\rho_0^{(\kappa)}(ll') \neq 0$, $\kappa = 0, 1, ...2l$. The time variation of the polarization moments of the ensemble of particles within the solid angle $d\Omega = \sin\beta \, d\beta \, d\gamma$ may be expressed as:

$$\begin{aligned} \frac{d}{dt} \overset{\approx}{\rho}{}_q^{(\kappa)}(ll') &= -\gamma_0(ll') \overset{\approx}{\rho}{}_q^{(\kappa)}(ll') + n_0 v f(\vec{v}) \sum_{\kappa,l,l',} \sigma_q^{\kappa\kappa_1}(ll'; l_1 l_1' | v) \overset{\approx}{\rho}{}_q^{(\kappa_1)}(l_1 l_1') \\ &+ n_0 v f(\vec{v}) \sigma_q^{(\kappa)}(ll'|v) \end{aligned}$$

where $\gamma_0(ll')$ is the radiative damping constant, $f(\vec{v})$ is the relative velocity distribution function of the colliding particles, $\sigma_q^{\kappa\kappa_1}(ll';l_1l_1'|v)$ is the matrix element of the collisional relaxation matrix of angular momenta of the hydrogen atom, and $\sigma_q^{\kappa}(ll'|v)$ is the matrix element of the collisional excitation matrix of the hydrogen atom to the state n=3 with the ordering of momenta in this state.

The contribution to the change of polarization moments $\overset{\approx}{\rho}{}_q^{(\kappa)}(ll')$ due to collisions taking place within the solid angle $d\omega$ and within the relative velocity interval $[\vec{v}, \vec{v}+d\vec{v}]$ may be expressed with the use of the rotation matrix $D_{q'q_1}^{(\kappa_1)*}(\alpha\beta\gamma)$ (Chapter 1) as follows:

$$\Delta \tilde{\rho}_q^{(\kappa)}(ll') = -\gamma_0(ll')\,\tilde{\rho}_q^{(\kappa)}(ll')$$

$$+n_0 v f(\vec{v})d\vec{v} \sum_{q_1 q' l_1 l_1' \kappa_1} D_{q_1 q}^{(\kappa)}(\alpha\beta\gamma) D_{q'q_1}^{(\kappa_1)*}(\alpha\beta\gamma)\sigma_{q_1}^{\kappa,\kappa_1}(ll';l_1l_1'|v)\,\tilde{\rho}_{q'}^{(\kappa_1)}(l_1l_1')$$

$$+n_0 v f(\vec{v})d\vec{v}\sum_{q_1} D_{q_1 q}^{x}(\alpha\beta\gamma)\sigma_{q_1}^{\kappa}(ll'|v)$$

The change of the polarization moments within the $\tilde{x}\,\tilde{y}\,\tilde{z}$ frame of reference under stationary conditions, making use of the axial symmetry of the collisional process, is given by the equation:

$$-\gamma_0(ll')\,\tilde{\rho}_q^{(\kappa)}(ll') + n_0\sum < v\sigma_q^{\kappa\kappa_1}(ll';.l_1l_1') > \tilde{\rho}_q^{\kappa}(l_1l_1') + n_0 N_q^{\kappa}(ll') = 0 \tag{7.13}$$

where $< v\sigma_q^{\kappa\kappa_1}(ll';l_1l_1') >$ is the matrix element of the rate constant matrix of the polarization moments. For an axially symmetric collisional process, we have the following expression for these matrix elements:

$$< v\sigma_q^{\kappa\kappa_1}(ll';l_1l_1') >= \sum_{q',L}(-1)^{q+q'}\begin{bmatrix} \kappa & \kappa_1 & L \\ q & -q & 0 \end{bmatrix}\begin{bmatrix} \kappa & \kappa_1 & L \\ q' & -q' & 0 \end{bmatrix}$$

$$\int d\vec{v} P_L(\cos\beta) v\sigma_{q'}^{\kappa\kappa_1}(ll';l_1l_1'|v)f(\vec{v}) \tag{7.14}$$

The collision excitation term at the left hand side of the equation (7.13) takes the form:

$$N_q^{\kappa}(ll') = \delta_{q0}\int P_q(\cos\beta)\sigma_q^{(\kappa)}(ll'|v)vf(\vec{v})d\vec{v} \tag{7.15}$$

For the case where the cross sections $\sigma_q^{\kappa\kappa_1}$ and σ_q^{κ} are dependent only on the absolute value of the relative velocity, the right hand sides of equations

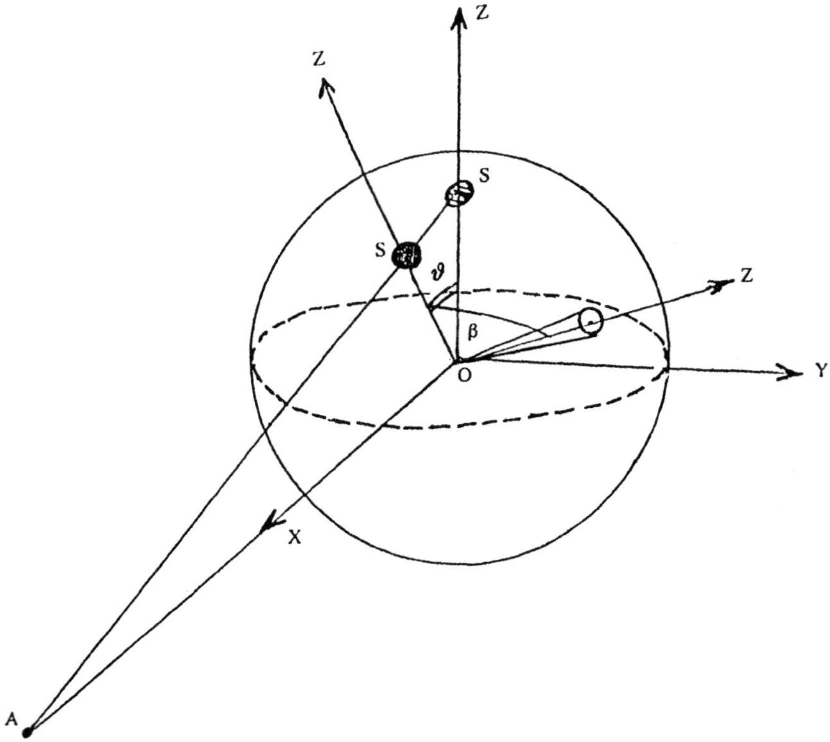

Figure 80. Coordinate system used for the description of spectropolarimetric effects in the solar chromosphere (O is the solar disc center, A is the ground based observer)

(7.14) and (7.15) are expressed in terms of the coefficients of expansion of the velocity distribution function within Legendre polynomials basis: :

$$< v\sigma_q^{\kappa\kappa_1}(ll';l_1l_1') >= \sum_{q',L}(-1)^{q+q'}\begin{bmatrix} \kappa & \kappa' & L \\ q & -q & 0 \end{bmatrix}\begin{bmatrix} \kappa & \kappa' & L \\ q' & -q' & 0 \end{bmatrix}\frac{2}{2L+1}$$

$$\int v^3\sigma_q^{\kappa\kappa_1}(ll';l_1l_1'|v)\, f_L(v)dv$$

$$N_q^{\kappa}(ll') = \delta_{q0}\frac{2}{2\,L+1}\int v^3\sigma_q^{(\kappa)}(ll'|v)\, f_L(v)dv$$

It is evident that in the frame of reference $\tilde{x}, \tilde{y}, \tilde{z}$, collisions induce population $\tilde{\rho}_0^{(0)}(ll)$ and alignment $\tilde{\rho}_0^{(2)}(ll)$, and if $l > 1$ the zero order components of polarization moments of the higher ranks $\tilde{\rho}_0^{(\kappa)}(ll), \kappa > 3$ may be also excited. Moreover, as collisions induce the coherent superposition of different values of the angular momentum of the excited hydrogen states, the non-diagonal by l polarization moments will also appear: $\tilde{\rho}_0^{(\kappa)}(ll')$. This is very important to the accounting for the mixing of polarization moments of different ranks due to perturbation, decreasing the symmetry of the collisional process. Such a decrease in symmetry may take place if there exist a local magnetic field non-parallel to the axis of collisional excitation, as well as when the non-homogeneities of the magnetic field are accounted for. The observer from Earth measures the degree of polarization in the frame of reference with OX axis directed towards Earth; $\tilde{OY}$ is the same as OY and the plane of the solar disc is ZY plane. Polarization moments within the frame of reference XYZ are expressed in terms of the quantities $\tilde{\rho}_q^{(\kappa)}(ll')$ with the help of the elements of the rotation matrix $D^\kappa(0,\vartheta,0)$ where ϑ is the angle $(\widehat{z\,\tilde{z}})$

$$
\begin{aligned}
\rho_0^0(ll') &= \tilde{\rho}_0^0(ll') \\
\rho_0^{(2)}(ll') &= D_{00}^{(2)}(0,\vartheta,0)\,\tilde{\rho}_0^{(2)}(ll') = \tfrac{1}{2}(3\cos^2\vartheta-1)\,\tilde{\rho}_0^{(2)}(ll') \\
\rho_1^{(2)}(ll') &= D_{01}^{(2)}(0,\vartheta,0)\,\tilde{\rho}_0^{(2)}(ll') = \tfrac{\sqrt{6}}{4}\sin 2\vartheta\,\tilde{\rho}_0^{(2)}(ll') \\
\rho_2^{(2)}(ll') &= D_{02}^{(2)}(0,\vartheta,0)\,\tilde{\rho}_0^{(2)}(ll') = \tfrac{\sqrt{6}}{4}\sin^2\vartheta\,\tilde{\rho}_0^{(2)}(ll')
\end{aligned}
\qquad (7.16)
$$

Choosing in the plane ZY two orthogonal directions determined by the polarization vectors $\vec{e}_{\parallel}$ and $\vec{e}_{\perp}$ ($\vec{e}_{\parallel}$ vector making α angle with OZ axis), in the n = 3 $\longrightarrow$ to n = 2 transition, the intensities of the vibrations polarized parallel to $\vec{e}_{\perp}$ and $\vec{e}_{\parallel}$ polarization vectors are given by

$$
\begin{aligned}
I_{\parallel} &= I_0 \textstyle\sum_{ll'}\{W_{ll}^{(0)}\rho_0^0(ll) + W_{ll'}^{(2)}[(1-3\cos^2\alpha)\rho_0^{(2)}(ll')/\sqrt{6} \\
&+ Re(i\sin 2\alpha\rho_1^{(2)}(ll')) + Re(\sin^2\alpha(ll'))]\} \\
I_{\perp} &= I_0 \textstyle\sum_{ll'}\{W_{ll}^{(0)}\rho_0^0(ll) + W_{ll'}^{(2)}[(1-3\sin\alpha^2)\rho_0^{(2)}(ll')/\sqrt{6} \\
&- Re(i\sin 2\alpha\rho_1^{(2)}(ll')) + Re(\cos\alpha^2(ll')\rho_2^{(2)}(ll'))]\}
\end{aligned}
\qquad (7.17)
$$

where

$$
W_{ll'}^{(x)} = \sum_{l_0} i^{l'-l}[2l_0||1||3l][2l_0||1||3l']\sqrt{(2l+1)(2l'+1)}\left\{\begin{matrix} 1 & 1 & x \\ l & l' & l_0 \end{matrix}\right\}
$$

$\left\{\begin{matrix} a & b & c \\ d & e & f \end{matrix}\right\}$ is the 6j symbol and $[2l_0 \,||\, 1 \,||\, 3l]$ is the reduced matrix element /16/.

Using (7.16) and (7.17) it is easy to obtain the expression for the degree of polarization of H_α line observed from the earth:

$$\mathcal{P} = \frac{\cos^2\vartheta \cos 2\alpha}{1/P - \sin^2\vartheta}. \tag{7.18}$$

Where P is the polarization degree calculated in the frame of reference connected to the symmetry axis of the collisional excitation process,

$$P = \frac{-3\sum W_{ll'}^{(2)}\,\tilde{\rho}_0^{(2)}(ll')}{\sum(2\sqrt{6}W_{ll}^{(0)}\rho_0^{(0)}(ll) - W_{ll'}^{(2)}\,\tilde{\rho}_0^{(2)}(ll'))}.$$

Formulae (7.13-7.15) show that the changes of the polarization moments due to collisions and as a result the degree of polarization of emission (7.18) are dependent on the relative velocity distribution function of the colliding particles. Ignoring the effect of transport on the particles velocity-distribution function, two kinetic models may be used to compute the degree of polarization of the H_α hydrogen line excited by proton impact.

Within the frame of the first model, the direct impact excitation of n=3 state is performed by fast projectiles with ordered velocity vectors (sect. 7.2.2). In this case the second term in the right hand side of the kinetic equation is zero and, in the Born approximation, the impact excitation cross section of the n=3 state from the ground level with the ordering of the angular momenta /14/ is given by:

$$\begin{aligned}\sigma_0^{(\kappa)}(ll'|v) &= 8\pi(\frac{e^2}{\hbar v})^2\sqrt{(2l+1)(2l'+1)}(-1)^{\kappa+l'}\begin{bmatrix} l & l' & \kappa \\ 0 & 0 & 0\end{bmatrix} \\ &\int\frac{dq}{q^3}[3l||j_l(qr)||10][3l'||j_{l'}(qr)||10]P_\kappa(\cos\beta)\end{aligned}$$

The angular dependence of the velocity distribution function is given by the formula (sect.7.2.2):

$$f(\vec{v}) = \left(v^4 + \frac{\varphi(v)}{\cos\beta}\right)^{-5/2}\delta(v - v_0)$$

where $\varphi(v)$ is a function weakly dependent on velocity.

The multipole moment of this function, taking into account that $\beta \in [0, \pi/2)$ is represented by:

$$f_\kappa(v) = \int_0^{\pi/2} v^3 f(\vec{v})P_\kappa(\cos\beta)\sin\beta d\beta dv$$

In order to illustrate the velocity dependence of these moments, the ratios f_2/f_0 and f_4/f_0 are given in Fig. 81 by dashed lines; the solid lines represents the results of calculations of the degree of polarization. The rapid

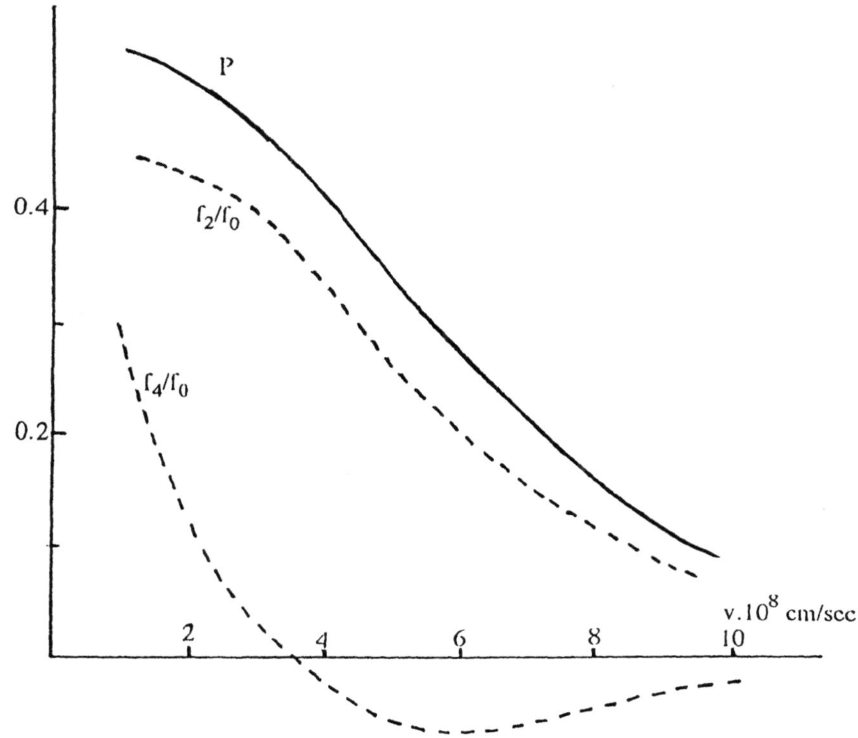

Figure 81. Velocity dependencies of the multipole momentum ratios and of the degree of polarization calculated for the case of direct proton impact alignment of hydrogen in the upper chromosphere

decrease of the polarization degree with velocity may be explained by the decrease of the multipole moment $f_2(v)$, because the corresponding velocity dependencies of the excitation and alignment cross sections are rather weak.

The second model incorporates the effect of the ordered collisional mixing of populations of the excited atomic states (Chapter 5). A drift velocity distribution function was used in this case:

$$f(\vec{v}, v_0) = (\gamma/\pi)^{3/2} \exp\left(-\gamma(v^2 - 2v_0 v \cos\theta + v_0^2)\right) \tag{7.19}$$

where

$$\gamma = \alpha\beta/(\alpha+\beta), \quad \alpha = \sqrt{m_a/2KT_a}, \quad \beta = \sqrt{m_b/2KT_b}.$$

Here m_a and m_b are the masses of the impeding proton and of the target atom.

Expanding the function (7.19) with respect to Legendre polynomials, and introducing the dimensionless parameter $\lambda = \sqrt{\gamma} v_0$, one arrives at the following expression for the multipole moments:

$$f_e(v,\lambda) = \frac{2l+1}{2\pi}\left(\frac{\lambda}{v_0}\right)^3 \sqrt{\frac{v_0}{v}} I_{l+1/2}(2\lambda^2 \frac{v}{v_0}) \exp\left\{-\lambda^2[1+(v/v_0)^2]\right\} \quad (7.20)$$

where $I_{l+1/2}$ is the Bessel function of purely imaginary argument. The parameter λ characterizes the anisotropy of the distribution. For $\lambda = 0$, all multipole moments except the moment of zero rank vanish and the velocity distribution becomes a Maxwellian one. For $\lambda = \infty$, the values of the multipole moments do not depend on l, and the distribution function (7.19) describes the collisions in the counterpropagating beams.

TABLE 7.1. H_α polarization computed for various proton energies and slit positions (see Fig. 77); observed polarization and proton energy deduced from the observations

v	E	Degree of polarization in %			
10^8 cm/sec	kev	Slit I	Slit II	Slit III	Slit IV
1	5.2	9.7	7.8	5.0	17.7
2	20.7	9.4	7.7	4.9	17.2
3	46.6	8.8	7.2	4.5	16.0
4	82.9	7.8	6.4	4.1	14.1
5	129	6.7	5.5	3.5	12.1
6	186	5.6	4.5	2.9	9.9
7	253	4.5	3.7	2.4	7.9
8	331	3.5	2.9	1.9	6.2
9	471	2.9	2.4	1.6	5.1
Observational data of the degree of polarization		4.5 ± 1.6	5.5 ± 1.7	3.5 ± 0.9	2.2 ± 0.9
Energy interval of protons (in kev)		(271 ± 98)	(136 ± 87)	(127 ± 73)	> 700

The polarization moments may be calculated within the frame of the second model using the equation (7.13) and, using (7.19) and (7.20), the last left hand side term may be expressed as follows:

$$N_q^\kappa(ll'|\lambda) = \delta_{q0}\frac{9}{8}e^{-\lambda^2}(\frac{e^2 q_0}{\hbar})^2\sqrt{\frac{\gamma}{\lambda}}\cdot\sqrt{(2l+1)(2l'+1)}(-1)^{l'+\kappa}\begin{bmatrix} l & l' & \kappa \\ 0 & 0 & 0 \end{bmatrix}$$

$$\int_0^\infty dx\sqrt{x}e^{-x^2}I_{\kappa+1/2}(2\lambda x)\int_{Q/x}^\infty [3l\|j_l\|10][3l'\|j_{l'}\|10]ds/s^3 \qquad (7.21)$$

where Q is a dimensionless parameter proportional to $\sqrt{\gamma}$.

The absence of a Legendre polynomial in the right hand side of (7.21) is explained by the predominance of scattering at small angles, and the infinite upper limit is taken in the integral because of the rapid decrease of the function under the integral. The second term at the left hand side of (7.13) was calculated within the impact parameter approximation using straight line trajectories and, assuming that the internuclear distance dependence of the interaction, follows a power law. The resulting system of nine equations has been numerically integrated within a wide range of impact parameters. As a result the right hand side of (7.14) may be written in a simple way:

$$< v\sigma_q^{\kappa\kappa'}(ll';l_1l_1') >= \overline{v}\sum_{q',L}(-1)^{q+q'}\begin{bmatrix} \kappa & \kappa' & 2L \\ q & -q & 0 \end{bmatrix}\begin{bmatrix} \kappa & \kappa' & 2L \\ q' & -q' & 0 \end{bmatrix}$$

$$\sigma_q^{\kappa\kappa'}(ll';l_1l_1'|\overline{v})F_{2L}(\lambda) \qquad (7.22)$$

where $\sigma_q^{\kappa\kappa'}(ll';l_1l_1'|\overline{v})$ is the cross-section of relaxation of angular moments in the counterpropagating beams, calculated at a relative velocity $\overline{v}$ equal to $1/\sqrt{\gamma}$ and where $F_l(\lambda)$ is the dynamic multipole moment:

$$F_l(\lambda) = \lambda^l e^{-\lambda^2}\frac{\Gamma(\frac{l}{2}+2)}{\Gamma(l+3/2)}\phi\left(\frac{l}{2}+1, l+\frac{3}{2};\lambda^2\right).$$

Here $\phi\left(\frac{l}{2}+1, l+\frac{3}{2};\lambda^2\right)$ is the degenerated hypergeometric function.

The system (7.22) was resolved for some values of the anisotropy parameter λ and for two values of the absolute temperature of the incident beam: $T_1 = 10^7$ K and $T_2 = 10^8$ K. As the temperature of the intermediary layer is about $10^4 K$ one may write $T_i > T$ (i=1, 2) and for both cases $\alpha \gg \beta_i$ (i=1, 2). Therefore, it was considered that $\gamma \approx \beta_i$ and that $\lambda = \sqrt{\gamma}v_0 \approx \sqrt{\beta}v_0$. .

The results of calculations are presented in Fig. 82. For $\lambda = 1$, the dependencies of $f_2(v,\lambda/f_0(v,\lambda)$ and $f_4(v.\lambda)/f_0(v,\lambda)$ on the dimensionless parameter $x = v/v_0$ were calculated by the formula (7.20). The drift velocities in this case were of the same order of magnitude as the thermal one. These velocity dependencies for the drift function (7.19) are opposite to those for the velocity distribution function examined in sect. 7.2.2. At large velocities, because of the asymptotic behaviour of Bessel functions, these ratios are proportional to (2l + 1) and do not depend on λ. The ratios $F_2(\lambda)/F_0(\lambda)$ and $F_4(\lambda)/F_0(\lambda)$ are also given by dashed lines. The

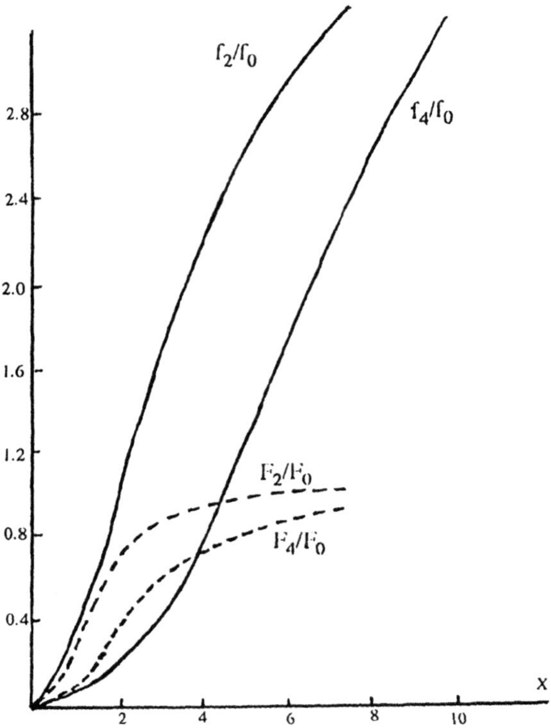

Figure 82. **Calculated dependencies of ratios of multipole moments and dynamic multipole moments on the dimensionless parameter $x = v/v_0$ for the drift velocity distribution function (7.19)**

calculated degree of polarization is presented in Fig. 83 as a function of the anisotropy parameter $\lambda = \sqrt{\gamma}/v_0$. Values of λ greater than 2 were not considered because plasma instabilities could occur for drift velocities of $(1.5 - 2)v_T$. The curves in this figure are practically equal at T greater than 10^7 K. Therefore, comparing the observed with the computed polarization degree of Fig. 83 for T higher than 10^7 K, one can see that an agreement is obtained for a drift velocity equal to $0.3\, v_T$.

Using the dependence of the degree of polarization on the projectile velocity (Figs. 81, 82), it is possible with the help of formula (7.18) to determine the velocity dependence of the degree of polarization measured by an observer on the earth. Results of calculations for four angles between the OZ and $\tilde{OZ}$ axis, ($\vartheta = 30°, 27°, 21.5°, 40°$) corresponding to the observation conditions (entrance slit positions 1-4) are presented in Table 7.1. With the help of the minimum and maximum observed values data of P, i.e. $P_{max} = P + \Delta P$ and $P_{min} = P - \Delta P$, we derived E_{max} and E_{min},

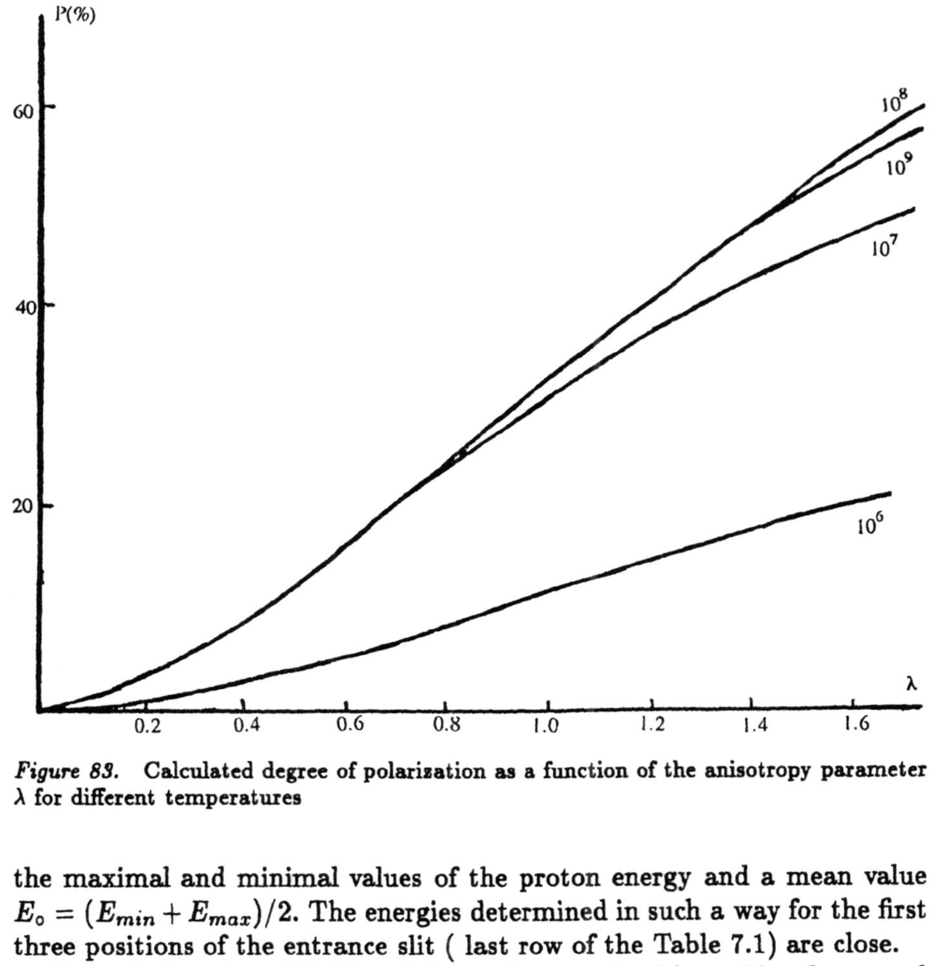

Figure 83. **Calculated degree of polarization as a function of the anisotropy parameter λ for different temperatures**

the maximal and minimal values of the proton energy and a mean value $E_0 = (E_{min} + E_{max})/2$. The energies determined in such a way for the first three positions of the entrance slit (last row of the Table 7.1) are close.

In analysing the reliability of the results one should consider the second Born approximation for the calculation of the cross sections /14/. In the case of small scattering angles and small transmitted pulses, the second Born correction for the direct and exchange scattering amplitude is less than for the first one, at least for the factor $k = P/\hbar$. The most adequate description for the studied energy range and small transmitted pulses is the Glauber approximation. However, this approximation is valid only when the transmitted kinetic moment is nearly orthogonal to the initial one. Uncertainties in the results may also be introduced by charge exchange collisions, which could induce an alignment of the hydrogen atoms. But for a target atom without an inner shell, the cross-section of such a process decreases rapidly at relative velocities higher than $e^2/\hbar^2$. For example, for proton-hydrogen scattering, the charge-exchange collision cross-section decreases as v^{-12}.

CONCLUSIONS

The spectropolarimetry offers, for the remote sensing of the structural features of a plasma entity, a tool which is complementary to all known spectroscopic methods. This technique is based on the fundamental physical concept of the polarization of an ensemble of atomic particles in ionized gases. The general theoretical relations between spectropolarimetric and kinetic parameters describing the particules velocity distribution, that are formulated mathematically making use of the irreducible representations of the three-dimensional rotation symmetry group, allow to derive these kinetic parameters from the measurement of the polarization of light. For this purpose, the principle integral relations connecting the observable polarization of radiation to the anisotropic pressure tensor of the particles responsible for the excitation of the emission spectrum have been derived and analysed in detail (see also /122/). As any spectroscopic technique, the spectropolarimetric diagnostics rely on the inversion of these integral relations.

The methods for determining the atomic parameters indispensable for the implementation of polarization spectroscopy have also been reviewed in detail. The benefits of the polarization spectroscopy technique have been tested by the derivation of the structural parameters of simple laboratory gas discharges, for which it was possible to make independent control comparative measurements. The most significant and promising application of the polarization spectroscopy of ionized gases as a remote sensing tool have been realized in solar astrophysics. Using space and ground based observations, polarization spectroscopy in the visible and UV provided a diagnosis of the presence of low energy protons accelerated in the solar corona during solar flares. Polarization spectroscopy will soon be widely used in various ground or space observatories for the study of the energy transport in solar flares and in other active astrophysical objects. The application of this technique in X-rays, shown to be very efficient for the study of the anisotropies in hot laboratory plasmas, is also very promising in astronomy.

The plasmas in the outer ionosphere of the Earth, and the plasma boundary sheaths are characterized by strong structural properties and anisotropies. They could be efficiently studied by the measurement of the polarization of the electromagnetic radiation they emit such as polar aurorae and twighlight emissions. In these cases also, physically significant applications of the polarization spectroscopy could be anticipated in the near future.

REFERENCES

Zeeman, P. (1897), *Phil. Mag.*, **43**, 226

Hale, G.E. (1908), *Astrophys. J.*, **28**, 100

Hanle, W. (1924), *Zs. Physik*, **30**, 93

Bommier, V., Leroy, J.L, Sahal-Brechot, S. (1986), *Astron.Astrophys.*, **156**, 90

Stenflo, J.O. (1994), *Solar Magnetic Fields*, Kluwer Academic Publishers, Dordrecht

Hyder, C.L. (1964), *Ph. D. Thesis*, Univ. Colorado

Hénoux, J.C., Chambe, G., Semel, M., Sahal-Bréchot, S., Woodgate, B., Shine, D., Beckers, J. and Machado, M. (1983), *Astrophys. J.*, **265**, 1066

Kieffer, J.C., Matte, J.P, Chaker, M., Beaudoin, Y., Chien, C.Y., Coe, S., Mourou, G., Dubau, J., Inal, M.K. (1993), *Phys. Rev. E*,**48**, 6, 4648

Sahal-Bréchot, S. (1984), *Ann. de Phys. (Paris)*, **9**, 705

Skinner, H.W.B. (1926), *Proc. Roy. Soc. (London)*, **112**, 642

Kazantsev, S.A. (1983), *Sov. Phys. Uspekhi.(USA)*, **26**, 328

Kazantsev, S.A. (1983), *JETP Letters. (USA)*, **37**, 159

Blum, K. (1981), *Density matrix theory and applications*, NY London Plenum Press

Landau, L.D., Lifshits, E.M.(1974), *Theoretical Physics 3. Quantum Mechanics*, Moscow Science 752p.

Veil, G. (1986), *Group theory and quantum mechanics*, Moscow Science, 495p.

Varshalovich, D.A., Moskalev, A.I., Khersonsky, V.E.(1975), *Quantum theory of momentum. St.Petersburg Science*, 439p.

Rebane, V.N.(1980), *Ph. D. Thesis*, Leningrad State University

Diakonov, M.I., Perel, V.I.(1964), *JETP (USSR)*, **47**, 1483

Omont, A.(1973), *Journ. Physique*, **34**, 179

Berestetsky, V.B., Lifshits, E.M., Pitaevsky, L.P.(1968),*Relative Quantum theory*, Part I, Moscow Science 480 p.

Diakonov, M.I.(1964), JETP (USSR). **47**, 2213

Chaika, M.P.(1975), *Interference of degenerated atomic states*, Leningrad State University Publ. 192p.

Fano, U., Macek, J.H.(1973), *Rev. Mod. Phys.*, **45**, 553

Frish, S.E.(1963), *Optical Spectra of Atoms*, Moscow Nauka, 640p

Frish, S.E.(1970), *Spectroscopy of Gas Discharge Plasma*, Leningrad (USSR) Science Publ.

Perel, V.I., Rogova, I.V.(1973), *JETP (USSR)*, **65**, 1012

Prilipko, V.K., Alexandrov, E.B.(1983), *Optics and Spectroscopy (USSR)*, **55**, 560

Golant, V.E., Zhilinsky, A.P., Sakharov, S.A.(1977), *Fundamentals of Plasma Physics*, NY, John Wiley & Sons 405p.

Shkarovsky, I.P., Johnston, T.W., Bachynsky, M.P.(1966), *The particle kinetics of plasmas*, London. Addison-Wesley Publ.

McFarland, R.H.(1964), *Phys. Rev.*, **133**, 986

McFarland, R.H., Soltysik, E.A. (1963), *Phys. Rev.*, **129**, 2581

Alexandrov, E.B. (1964), *Optics and Spectroscopy (USSR)*, **16**, 377

Anderson, N, Gallagher, J.W., Hertel, I.V. (1988), *Phys. Rep*, **165**, 1

Drukarev, G.F. (1978), *Collisions of electrons with atoms and molecules*, Moscow Science, 255p.

Burke, P.G., Robb, W.D. (1975), *Adv. Atom. Molec. Phys.*, **11**, 143

Drachev, A.I., Kazantsev, S.A., Subbotenko, A.V.(1990), *Optics and Spectroscopy (USA)*, **69**, 585

Kazantsev, S.A., Polynovskaya, N.Ya., Piatnitsky, L.N., Edelman S.A. (1988), *Sov. Phys. Usp. (USA)*, **31**, 785

Hénoux, J.-C., Chambe, G. (1990), *JQSRT*, **44**, 193

Kazantsev, S.A., Firstova, N.M., Gubin A.V., Lankevich N.A. (1991), *Optics and Spectroscopy (USA)*, **70**, 580
Pool, Ch. (1970), *EPR Spectroscopy*, Moscow Mir 557p.
Griem, H.R.(1964), *Plasma spectroscopy*, New York, McGraw Hill, 580p.
Kazantsev, S.A., Subbotenko, A.V. (1986), *Optics and Spectroscopy (USA)*, **60**, 157
Kazantsev, S.A., Subbotenko, A.V.(1987), *Leningrad State University Vestnik (USSR)*, **25**, 20
Larkina, L.T, Engelsht, V.S.(1973), *All-Union Information Institute Proceedings (USSR)*, 6917-73, 94p
Preobrazhensky, N.G.(1976), *Ill-posed reversed problems of atomic physics*, Novosibirsk Science 133p
Lochte-Holtgreven, W.(1968), Plasma Diagnostics. Amsterdam North Holland
Kazantsev, S.A., Kisling, A., Chaika, M.P.(1973), *Optics and Spectroscopy (USA)*, **34**, 714
Kazantsev, S.A., Markov, V.P. (1974), *Optics and Spectroscopy (USA)*, **36**, 355
Kazantsev, S.A., Kisling, A., Chaika, M.P.(1974), *Optics and Spectroscopy (USA)*, **36**, 606
Kazantsev, S.A., Kisling, A., Chaika, M.P.(1975), *Leningrad State University Vestnik (USSR)*, **10**, 33
Kazantsev, S.A., Rys, A.G., Chaika, M.P.(1978), *Optics and Spectroscopy (USA)*, **44**, 249
Kazantsev, S.A., Markov, V.P., Morozova S.L., Chaika, M.P.(1979), *Optics and Spectroscopy (USA)*, **46**, 619
Kazantsev, S.A., Polzik, E.S. (1976), *Optics and Spectroscopy (USA)*, **41**, 645
Kazantsev, S.A., Rys, A.G.(1977), *Optics and Spectroscopy (USA)*, **43**, 339
Kazantsev, S.A. (1980), *Leningrad State University Vestnik*, **4**, 52
Kazantsev, S.A., Eiduk, V.I.(1978), *Optics and Spectroscopy (USA)*, **45**, 735
Kazantsev, S.A., Markov, V.P. (1979), *Optics and Spectroscopy (USA)*, **47**, 112
Kazantsev, S.A., Rys, A.G., Chaika, M.P.(1983), *Optics and Spectroscopy (USA)*. **54**, 124
Kazantsev, S.A., Rys, A.G.(1989), *Leningrad State University Vestnik (USSR)*, **4**, 24
Tsendin, L.D., Golubovsky, Yu.B. (1977), *Sov. J. Techn. Phys. (USSR)*, **47**, 1839
Kazantsev, S.A.(1982), *Optics and Spectroscopy (USA)*, **52**, 559
Drachev, A.I., Kazantsev, S.A., Subbotenko, A.V.(1991), *Optics and Spectroscopy (USA)*, **70**, 159
Kazantsev, S.A., Rys, A.G., Subbotenko, A.V.(1988), *Sov.J.Plasma Physics (USA)*, **14**, 143
Grigorieva, V.N., Kazantsev, S.A., Kudriashova, M.A.(1983), *Optics and Spectroscopy (USA)*, **54**, 247
Kazantsev, S.A., Subbotenko, A.V.(1984), *Sov. J. Plasma Physics (USA)*, **10**, 78
Kazantsev, S.A., Svelokuzov, A.E., Subbotenko, A.V. (1986), *Sov. J. Techn. Phys.*, **56**, 1091
Kazantsev, S.A., Smirnova, O.M., Subbotenko, A.V. (1988), *Leningrad State University Vestnik (USSR)*, **18**, 9
Kazantsev, S.A., Subbotenko, A.V.(1987), *Journ. Phys. Ser. D.*, **20**, 741
Fermi, E.(1949), *Phys. Rev.*, **75**, 1169
Kuzovnikov, A.A., Savinov, V.P.(1973), *Moscow State University Vestnik (USSR)*, **2**, 215
Kuzovnikov, A.A., Savinov, V.P.(1973), *Radiotechnics and Electronics (USSR)*, **18**, 816
Kazantsev, S.A., Subbotenko, A.V.(1984), *Sov. J. Techn. Phys. Lett.*, **10**, 1251
Drachev, A.I., Kazantsev, S.A., Subbotenko, A.V.(1991), *Optics and Spectroscopy (USA)*, **71**, 527
Drachev, A.I.,Kazantsev, S.A., Rys A.G., Subbotenko, A.V.(1988), *Optics and Spectroscopy (USA)*, **64**, 706
Petrashen, A.G., Rebane, V.N., Rebane, T.K.(1984), *JETP (USSR)*, **89**, 147
Petrashen, A.G., Rebane, V.N., Rebane, T.K.(1985), *Optics and Spectroscopy (USA)*,

58, 481

Kazantsev S.A., Petrashen A.G., Polezhaeva N.T., Rebane V.N., Rebane T.K.: (1990), *Optics and Spectroscopy (USA)*, **68**, 544

Kazantsev S.A., Petrashen A.G., Polezhaeva N.T., Rebane V.N., Rebane T.K.: (1990), *Optics and Spectroscopy (USA)*, **69**, 582

Kazantsev, S.A., Petrashen A.G., Polezhaeva, N.T., Rebane, V.N., Rebane T.K. (1987), **JETP Letters (USA)**, **45**, 18

Kazantsev, S.A., Polezhaeva, N.T., Rebane, V.N.(1987), *Optics and Spectroscopy (USA)*, **63**, 15

Kazantsev S.A.,Petrashen A.G., Polezhaeva N.T, Rebane V.N. (1989), *Optics and Spectroscopy (USA)*, **66**, 309

Kazantsev S.A., Petrashen A.G, Polezhaeva N.T., Rebane V.N.:(1990), *Optics and Spectroscopy (USA)*, **68**, 740.

Kazantsev, S.A., Polynovskaya, N.Ya, Piatnitsky, L.N., Edelman S.A. (1985), *Optics and Spectroscopy (USSR)*, **58**, 28

Gorny, M.B., Kazantsev, S.A., Matisov, B.G.(1985), *Zs. Physik*, **322**, 25

Unsold, A.(1949), *Physics of Stellar Atmospheres*, Springer

Shklovsky, I.S.(1962), *Physics of the Solar Corona*, Moscow Nauka

Ohman, Y.(1929), *Monthly Notices Roy. Astron. Soc.*, **89**, 479

Zanstra, H.(1950), *Monthly Notices Roy. Astron. Soc.*, **110**, 491

Thiessen, G.(1951), *Zs. Astrophys.*, **30**, 8

Hyder, C.L. (1964), *Astrophys J.*, **140**, 817

Hyder, C.L. (1965), *Astrophys. J.*, **141**, 1374

Warwick, J.M., Hyder, C.L. (1965), *Astrophys. J.*, **141**, 1362

Nikolski, G.M., Khetsuriani, T.S. (1970), *Sov. Astron.*, **13**, 815

Leroy, J.L., Ratie, G., Bommier, V. (1977),*Astron. Astrophys.*, **54**, 811

Heasley, J.M. et al.(1974), *Astrophys. Journ.*, **192**, 181

Bommier V.(1980), *Astron. Astrophys.*, **87**, 109

Sahal-Brechot, S. et al.(1977), *Astron. Astrophys.*, **59**, 223

House, L.L., Cohen, L.C. (1969), *Astrophys. Journ.*, **157**, 216

Charvin, P. (1965), *Ann. Astrophys*, **28**, 877

Hyder, C.L. (1965), *Astrophys. J.*, **141**, 1382

Hyder C.L. et al. (1968),*Astrophys. J.*, **154**, 1039

Sahal-Brechot, S. (1974), *Astron. Astrophys.*, **36**, 335

Stenflo, J.O. (1976), *Astron. Astrophys.*, **46**, 61

Stenflo, J.O., Stenholm L. (1976), *Astron. Astrophys.*, **46**, 69

Stenflo, J.O. (1978), *Astron. Astrophys.*, **66**, 241

Svestka, Z. (1976), *Solar Flares*, Reidel, Dordrecht

Zirin, H.(1988), *Astrophysics of the Sun*, Cambridge University Press, Cambridge

Tandberg-Hanssen, and Emslie, A.G, (1988), *The physics of solar flares*, Cambridge University Press, Cambridge

Somov, B.V., (1992), *Physical processes in solar flares*, Kluwer Academic Publishers, Dordrecht

Hénoux, J.-C., Heristchi, D., Chambe, G., Machado, M., Woodgate, B., Shine, R. and Beckers, J.; (1983), *Astrophys. J.*, **119**, 233

Hénoux, J.-C. (1991), *Solar Polarimetry*, Proceedings of the Eleventh National Solar Observatory / Sacramento Peak Summer Workshop, ed. L.J. November

Hénoux, J.-C., Chambe,, G., Smith D., Tamres, D., Feautrier, N., Rovira, M., Sahal Bréchot, S. (1990), *Astrophys. J. Supplt.*, **73**, 303

Hénoux, J.-C., Firstova, N.M., Kazantsev S.A., Petrashen A.G. (1995), *Optics and Spectroscopy (USA)*, **78**, No 4

Syms, R.F., McDowell, M.R.C., Morgan, L.A., and Myerscough,, V.P. (1975), *J. Phys. B, Atom. Mol. Phys.*, Vol.8, No.17

Kleinpopen, H., Kruger, H. and Ulmer, R. (1962), *Phys. Letters*, **2**, 78

Kleinpopen, H., Kraiss, E. (1968), *Phys. Rev. Letters*, **20**, 361

Mahan, A.H.(1974), *Ph.D.thesis*, University of Colorado

Trubnikov, B.A. (1965), *Rev. Plasma Phys.*, **1**, 105

Machado, M.E., Avrett, E.H. et al. (1980), *Astrophys. J.*, **242**, 336

Hénoux, J.-C., Firstova, N.M., Kazantsev S.A., Petrashen A.G. (1993), *Optics and Spectroscopy (USA)*, **75**, 382

Hénoux J.-C.,Kazantsev S.A., Petrashen A.G., Firstova N.M.: (1993), *Europhys. Confer. Abstr.*, EGAS-25, Caen, **17D**, P2-98.

Kazantsev S.A., Feautrier N., J. C. Hénoux, Liaptsev A.V., Luchinkina V.V. (1995), *Astron. Astrophys. Rev.*, **6**

GPSR Compliance

The European Union's (EU) General Product Safety Regulation (GPSR) is a set of rules that requires consumer products to be safe and our obligations to ensure this.

If you have any concerns about our products, you can contact us on ProductSafety@springernature.com

In case Publisher is established outside the EU, the EU authorized representative is:

Springer Nature Customer Service Center GmbH
Europaplatz 3
69115 Heidelberg, Germany

Batch number: 10583998

Printed by Printforce, the Netherlands